T0360851

Industry Emergence

This book develops a novel industry emergence framework to explain the features, interaction, and synchronization of key elements for the birth and growth of new industries.

Organized around seven elements—firm strategy, technology, investment, supply networks, production, markets, and government—Theyel's framework provides inventors, managers, investors, scholars, and policymakers with a comprehensive understanding of how industries emerge, helping them to be more successful at influencing the birth and growth of new industries. Understanding industry emergence is important because new industries can offer the advancement of technology, improvements in human health and the environment, growth of firms, creation of jobs, and economic development.

With learning objectives, theories, tools, case studies, and end-of-chapter exercises, *Industry Emergence* will be a useful resource for students and professionals in engineering, science, business, and policy.

Gregory Theyel is a Professor at California State University, USA. He is also president of Green Visions, which assists companies with technology commercialization, and governments with the growth of new industries and economic development.

Industry Emergence

Strategic Management and Synchronization for New Industries

Gregory Theyel

NEW YORK AND LONDON

First published 2017
by Routledge
711 Third Avenue, New York, NY 10017

and by Routledge
2 Park Square, Milton Park, Abingdon, Oxon OX14 4RN

Routledge is an imprint of the Taylor & Francis Group, an informa business

Library of Congress Cataloging-in-Publication Data
A catalog record for this book has been requested

ISBN: 978-0-415-73499-8 (hbk)
ISBN: 978-1-315-81948-8 (ebk)

Typeset in Times New Roman
by codeMantra

Contents

List of Figures vi
Preface vii
Acknowledgments x

1 Understanding Industry Emergence 1
2 Firm Strategy 11
3 Technology 22
4 Investment 32
5 Supply Networks 43
6 Production 53
7 Markets 62
8 Government 74
9 Synchronization for Industry Emergence 84
10 Strategic Management for Industry Emergence 98

Index 103

List of Figures

1.1	Framework for Understanding Industry Emergence	5
2.1	Firm Strategy and Industry Emergence	13
2.2	UAV-Related Publications	14
3.1	Technology and Industry Emergence	24
3.2	Growth of Wearable Technology	26
3.3	Technology and Regenerative Medicine Industry Emergence	29
4.1	Investment and Industry Emergence	33
5.1	Supply Networks and Industry Emergence	45
5.2	Li-Ion Pricing and Energy Density	48
5.3	Global Wind Power Capacity	50
6.1	Production and Industry Emergence	54
7.1	Markets and Industry Emergence	64
7.2	Global Solar PV Electricity Generation	70
8.1	Government and Industry Emergence	75
8.2	Virtual Reality Patents	79
8.3	Global Biotechnology Sales Growth	81
9.1	Number of Firms (Percentage Growth)	87
9.2	Technology – Patenting (Percentage Growth)	88
9.3	Investment (Percentage Growth)	89
9.4	Supply Networks (Price Change of Inputs)	90
9.5	Production (Scale)	91
9.6	Markets (Percentage Growth)	91
9.7	Government (Spending)	92
9.8	Interaction of Elements for Synchronization	93

Preface

Industry emergence is the birth and early growth of new industries shaped by the interaction and synchronization of the strategic behavior of firms, technological innovation, investment, supply networks, production, markets, and the influence of government and other institutions. We use this description to guide our development of a framework and our understanding of the elements that drive the growth of new industries.

Emerging industries are groups of companies developing around a new, often disruptive idea, technology, product, or service. The purpose of this book is to provide inventors, firms, investors, scholars, and policymakers with understanding about how industries emerge and enable them to be more successful at influencing the birth and growth of new industries. Inventors and firms can benefit from a better understanding of the multiple elements influencing the progress of their technology and business. Investors can make better-informed decisions about their funding of companies. Scholars can present their research in a more systemic context, and policymakers can coordinate their efforts in concert with technology, markets, investment, and firm strategy. Understanding industry emergence is important because new industries can offer the advancement of technology, improvements in human health and the environment, growth of firms, creation of jobs, and economic development.

The unit of analysis for this book is the industry. This book is not specifically about emerging technologies, firms, markets, or countries. While all of these topics are related to emerging industries, this book's contribution is a better understanding of how industries emerge.

The book is organized around seven elements—firm strategy, technology, investment, supply networks, production, markets, and government. The order of the elements is not meant to imply an order of importance. Instead, different elements are more or less important depending on the characteristics of specific emerging industries. This is made clear through the many emerging industry case studies presented throughout the book.

Chapter 1 describes emerging industries, presents a framework of elements for explaining industry emergence, and introduces the concepts of synchronization and states of industry emergence. Synchronization is the coordination of multiple elements so that they reach a particular state simultaneously and is the key concept for industry emergence because important elements need to be synchronized in order for an industry to emerge.

Chapter 2 presents firm strategy, which is how a firm positions itself in an industry and how it uses resources to support its position. An industry consists of (many) firms and the strategic decisions of individual entrepreneurs/firms usually start the process of industry emergence.

Chapter 3 shows that technology, which is made up of design aspects, knowledge, and usually physical components, is often the seed for industry emergence. However, counter to what many scientists and engineers will tell you, emerging industries are not only driven by technology.

Chapter 4 presents how investment plays a critical role in the emergence of new industries because it facilitates the launching of firms, advancement of technology, refinement of production capabilities and supply networks, and development of markets. Investment is probably the most talked about element of emerging industries, but while it is important, the other elements in this book are also important because, for example, failing technology or market gaps, incompetent firm strategy, or misdirected government policy cannot be corrected solely with more investment.

Chapter 5 is about supply networks, which we describe as groups of individuals, firms, and institutions providing inputs for the development of a product or service, forming and growing as an industry emerges. The strongest supply networks encompass a wide variety of support industries, which aids in the development of alliances and benefits emerging industries, facilitating innovation and commercialization of technology.

Chapter 6 presents production, which plays an essential role for industry emergence, because technology remains a curiosity if we lack the ability to manufacture it into a product. We define production as involving the translation of ideas into reality, and production often facilitates the progression from invention to a commercialized product, leading to the emergence of an industry.

Chapter 7 describes how visionaries, early adopters, and eventually followers build markets for new industries. The rate of adoption of a technology depends on features of both the technology and the potential market.

Chapter 8 shows how government plays a special role in influencing all the other elements in the book. It is this omnipresent influence on all the other elements that makes government so special, and is its unique characteristic.

Chapter 9 focuses on the interaction of the elements influencing industry emergence described in the preceding chapters and elaborates on the concept of synchronization and its importance for the emergence of new industries. Element interaction and synchronization are related concepts because the interaction of the elements can result in co-development and coordination leading to synchronization. This chapter also elaborates on our three states of industry emergence – Concept, Validation, and Diffusion.

Chapter 10 presents how this book can provide inventors, firms, investors, scholars, and policymakers with understanding about how industries emerge and enable them to be more successful at influencing the birth and growth of new industries. We envision this book helping multiple parties as they play their roles in the emergence of an industry, but we also seek to continue the conversation about how to better understand and encourage the emergence of new industries.

Acknowledgments

This book began as a program on Emerging Industries funded by the UK Engineering and Physical Sciences Research Council (EPSRC) and the Gatsby Foundation. The program was led by Sir Michael Gregory, the Head of the Institute for Manufacturing (IfM) in the Department of Engineering at the University of Cambridge. I thank Professor Gregory for the opportunity he gave me to work on this research project and learn about this fascinating topic. Paul Heffernan was an early collaborator and leader, and he participated in many engaging conversations about the ideas in this book, and it would not have developed the way it has without his contribution. Many researchers (David Probert, Rob Phaal, Eoin O'Sullivan, Michelle Routley, Simon Ford, Tim Minshall, Nicky Dee, Finbarr Livesey, Laure Dodin, James Moultrie, Alex Driver, Ken Platts, Sirirat Lim, Jag Srai, David Kirkwood, Carol Walton, and Daniela Manca) were part of the IfM team participating in stimulating conversations and contributing significant insights that helped shape this book. My editor at Routledge Press, Sharon Golan, has been extremely supportive throughout the process of writing this book as she showed great confidence and belief in me and the importance of writing this book.

I have taught the material in this book to over 1,000 students, and their questions, research, and case studies have helped me improve the focus and explanation of the book's concepts and the relevance of the case study industries. Jordan Theyel provided valuable editing of multiple drafts, up-to-date, first-hand insight on many of the case studies, and useful ideas for improving the accessibility of the concepts. Marshall Theyel helped me with data analysis and mapping, particularly for the biotechnology and regenerative medicine case studies. Elizabeth Garnsey offered valuable insight and encouragement throughout the process suggesting literature and conceptualization of systems, evolutionary economics, and firm growth. Kay Hofmann read the entire book and offered valuable insight throughout. Allison Lau helped research and write the wearable healthcare devices case study. Amanda Gresham helped research and write the electric vehicle and social networking case studies. Adam Zwicker offered a valuable student's perspective on how the ideas throughout the book could be communicated more clearly and effectively. Finally, I acknowledge my wife, Nelli Theyel, who made writing this book much easier with her clear and helpful insight throughout the entire process, and who makes my life much happier and fulfilling.

1 Understanding Industry Emergence

Learning Objectives:

- Describe emerging industries
- Identify key elements affecting industry emergence
- Understand element synchronization
- Distinguish states of industry emergence

Key Concepts:

- Industry emergence
- Elements affecting industry emergence
- Synchronization
- States of industry emergence

1.1 Introduction

Emerging industries are groups of companies developing around a new, often disruptive idea, technology, product, or service. But why do industries emerge at such different rates? For example, why has the digital camera industry emerged since the mid-1990s to close to complete dominance today, while the solar photovoltaic industry is contributing only about 5 percent of electric power generated after more than 50 years of growth? The answer to this question is complex, just as is the process of industry emergence. This book explains how industries emerge when key elements such as firm strategy, technology, investment, supply networks, production, markets, and government interact and synchronize. **Synchronization**, or the coordination of multiple elements so that they reach a particular state simultaneously, is the key concept for industry emergence, because just as a conductor keeps an orchestra synchronized or as synchronized swimmers are coordinated in time, so too must important elements be synchronized in order for an industry to emerge.

The emergence of new industries is a complex and dynamic process, and it is vital because of the potential it offers for the advancement of technology, improvements in human health and the environment, growth of firms, creation of jobs, and economic development. Understanding how industries emerge is informative for inventors, firms, investors, and policymakers. Inventors and firms can benefit from a better understanding of the multiple elements influencing the progress of their technology. Investors can make better-informed decisions about their funding of companies, and policymakers can coordinate their efforts with technology, investment, markets, and firm strategy.

This book draws on multiple disciplines to illuminate industry emergence and present a framework of elements that influence how industries emerge. It shows the synchronization of the elements and the states of industry emergence and uses many case studies of emerging industries to support the usefulness of the framework. The unit of analysis for the book is the industry. This book is not specifically about emerging technologies, firms, markets, or countries. While all of these topics can be related to emerging industries, this book's contribution is a better understanding of how industries emerge. This chapter describes and explains emerging industries, presents a framework of elements for explaining industry emergence, and introduces the concepts of synchronization and states of industry emergence.

1.2 Industries and Emerging Industries

This chapter and book present industry emergence and its special nature and the interaction, co-evolution, and synchronization of important elements affecting new industries. But first, it is necessary to define industries and describe their growth in general.

1.2.1 Industries and their Growth

An **industry** is a group of firms aiming to meet the needs of a target group of customers with similar products and/or services. Within an industry, firms choose from multiple strategic positions. Firms can offer cost-leading products or services, or they can differentiate with luxury, or highly innovative products or services. Firms may have similar offerings or different offerings that provide similar utility, for example renewable energy firms may offer the same technology or a different technology that still generates electricity. Firms may choose to focus on different types of value chain activities. For example, some firms develop and produce products, while others market and sell them. A firm is described as vertically integrated if it is involved in all industry value chain activities. Unless a firm is vertically integrated, it relies on other companies for complementary resources to deliver a product or service. Firms' roles change depending on where they are in an industry value chain. Firms that develop and produce products have suppliers that provide raw materials. They are in turn considered suppliers to their customers who sell their products. These customers sell their products to end users, who are their customers. Thus, an industry is a complex system of firms, with each contributing to meeting the needs of a target group of customers.

Many authors have suggested patterns for the growth of industries.[1,2] Common approaches suggest that there is an initial phase where technology develops and firms vie for leadership. New ideas are explored, developed, and experimented with in order to see the direction of the technology and market. Over time, a standard is approached and there is what is referred to as a 'shakeout' of the firms that do not win the technology/product/business model standards battle and go out of business or are acquired. The winners with their standard setting technology, product, and/or business model lead an industry takeoff, or high growth period. With fewer competing firms and a growing market, sales usually increase, as do investment and profits. This is when production turns its focus to scaling for greater efficiency, supply chains develop, and the industry moves toward maturity. Decline and potentially rebirth follow industry maturity. This briefly describes the lifecycle for industries in general. The next section shifts our focus to industry emergence, where it will remain for the rest of the book.

1.2.2 Emerging Industries

Emerging industries are groups of companies developing around a new, often disruptive idea, technology, product, or service. They are new industries offering utility through new technology, products, or services to a group of customers. They often have new value chains and supply networks and are driven by a "disruptive idea (or convergence of ideas)"[3] that turns into a new product or service and a new industry. New industries emerge when entrepreneurs succeed in mobilizing resources in response to perceived opportunities.[4] These perceived opportunities motivate individuals and firms to focus their limited resources in a particular direction. These risk-takers hope that their hunch about an opportunity and the quality of their resources will be the right match for success.

The emergence of a new industry can be seen as a process that is based on a technological innovation meeting a new or existing customer need.[5]

A technological breakthrough is often responsible for a new industry. However, this breakthrough needs to meet the new or existing customers' needs better than alternatives, usually both on utility for customers and on price.

Emerging industries are said to be characterized by "novel and coherent structures, patterns and properties driving the process of self-organization in complex systems,"[6] and are described as "newly formed industries that have been created by technological innovations, shifts in relative cost relationships, emergence of new customer needs, or other economic and sociological changes that elevate a new product or service to the level of a potentially viable business opportunity."[7]

Common elements in the descriptions of industry emergence include technological innovation, market needs and opportunities, entrepreneurs, and complex systems. These descriptions imply interaction, co-development, and synchronization of the elements. Emerging industries are complex, and understanding them calls for an organized and systemic approach. Therefore, we integrate the descriptions above and present industry emergence as **the birth and early growth of new industries shaped by the interaction and synchronization of the strategic behavior of firms, technological innovation, investment, supply networks, production, markets, and the influence of government and other institutions**. We use this description to guide our development of a framework and our understanding of the elements that drive the growth of new industries. In the next section we present our framework, examining the interactions and synchronization of critical elements for industry emergence, with the help of example emerging industries.

1.3 Framework for Understanding Industry Emergence

The seven elements—firm strategy, technology, investment, supply networks, production, markets, and government—that we represent as horizontal 'process bands' in Figure 1.1 show the changes that generally occur as an industry emerges. The order of the elements is not meant to imply an order of importance. Instead, different elements are more or less important depending on the characteristics of specific emerging industries. This is made clear through the many emerging industry case studies presented throughout this book.

Vertically, we show three states of emergence: Concept, Validation, and Diffusion. The elements must be synchronized in order for an industry to grow from one state of emergence to the next. The seven elements and three states of emergence are based on the contributions of many

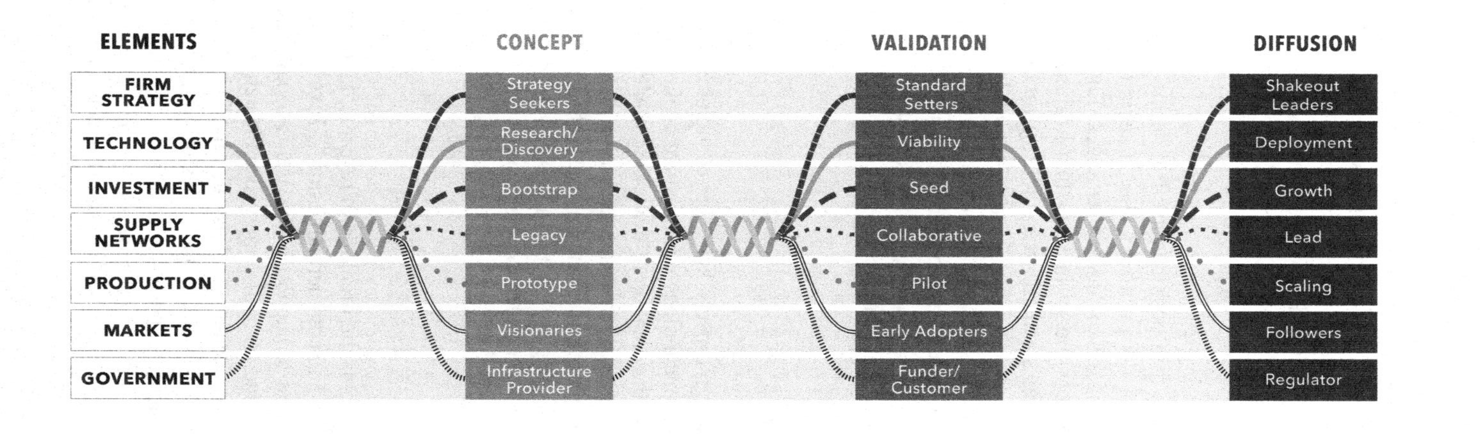

Figure 1.1 Framework for Understanding Industry Emergence.

authors. In the following section, we summarize an array of insights from numerous disciplines, including engineering, management, economics, and political science, in order to assemble our interdisciplinary framework for understanding industry emergence.

1.3.1 Firm Strategy

Firm strategy is how a firm positions itself in an industry and how it uses resources to support its position.[8] The strategic options, or positions, for firms change as their industry emerges. Early entrant firms search for opportunities while an industry first begins to emerge, followed by firms that set the standards in an industry, for example, Intel in the microprocessor industry and Ford in the automobile industry. Firms begin to compete for market share as an industry emerges and grows. As industry emergence continues, shakeout occurs as new industry entrants join in and others exit or are acquired, as the focus shifts from product to process innovation, and firms that excel at operational effectiveness lead the industry (see Figure 1.1).[9]

1.3.2 Technology

Technology entails design aspects, knowledge, and often physical components. The process of technology development begins with a research discovery, followed by the creation of a working model of the technology, that is, proof of the viability of the technology, and then advancement to deployment, which is the functional use of the technology (see Figure 1.1). Gaps in this progression may slow or even prevent the emergence of an industry, as some technologies do not advance to viability and deployment.

1.3.3 Investment

Investment often represents the lifeblood of research, innovation, development of value chains, and commercialization.[10] **Investment**, or financial resources committed to support a venture, is usually essential for an industry to emerge, and with this emergence the nature and extent of investment opportunities change. As illustrated in Figure 1.1, the availability and sources of investment are likely to vary over time as the other elements that affect emerging industries change.

Entrepreneurs often provide their own early financial resources, in what is called "bootstrapping." As an industry continues to emerge, risks often decrease or at least become better understood, so government grants and investors with profit motivation are more likely to be attracted. Investments from private ("angel") investors and venture capitalists often replace personal investment. During the later stages of industry emergence, firms may reach the point where their financing is internally generated in the form of retained earnings and sale of assets, or via the public offering of ownership, though even at this point, new entrants and those introducing radical innovations are likely to continue to require founder, government, venture capital, and other funding sources.[11]

1.3.4 Supply Networks

Supply networks transform to support the birth and growth of new industries. New industries are seldom formed around a single technology or by a single entity. Instead, multiple technologies and alliances are developed in new ways to meet new needs. **Supply networks**, or groups of individuals, firms, and institutions working together for a common purpose, form and grow as an industry emerges in order to deliver the technology, quality, and scale necessary to support a growing industry. As the first firms in a new industry piece together the components they need to develop prototypes, supply networks are usually sparse. Supply networks expand and collaboration builds in order for industries to continue to emerge, and eventually there is network consolidation around the leading technology and component suppliers (see Figure 1.1). Supply networks are not independent of the other elements in the emerging industries framework. They depend on investment in technologies and processes, which is influenced by market growth, firm strategy, and the development of production.

1.3.5 Production

Production, or the action of making or manufacturing goods from components or raw materials, also changes as an industry emerges (see Figure 1.1). Production is essential to be able to transform ideas or research discoveries into tangible products, and firms usually need to develop a broader range of production capabilities, as an industry emerges, in order to make their products. Early prototyping is essential for a concept or research finding to be advanced, and being able to make a new product is often necessary for markets to form and investments to be attracted. Pilot production enables testing and refinement of products and markets, and scaling production is usually critical for products and services to compete on price with the existing dominant products or services.

1.3.6 Markets

The development of **markets** for the products and services of emerging industries often follows a regular pattern from visionaries, to early adopters, to followers. The adoption of a new product or service depends upon features of both the technology and the potential customers.[12] The chief characteristic by which markets are categorized is the level of adoption. Markets begin with "visionaries" who see the possibility of a product or service, followed by people who are keen to adopt products or services as soon as possible (early adopters). Followers are next and are necessary for the emergence of an industry, as in many cases a product or service needs to reach a level of adoption in order for users to appreciate its value, for example, a critical mass of telephone users is necessary in order for there to be other people to phone.[13] The products or services of new industries often compete against the existing dominant products or services of the incumbent industry, for example gasoline-propelled vehicles vs. electric motor-propelled vehicles, in a battle that begins with visionary pioneers, leads to early adopters, and grows due to followers (see Figure 1.1).

1.3.7 Government

Government, or the governing entity of a country, state, or community, uses a range of mechanisms to influence the direction and pace of industry growth, and, as shown in Figure 1.1, tends to use them at different times for different effects. In the beginning of an industry's existence, the role of the government is most likely to involve providing the infrastructure for the development of technology and markets. As industries continue to emerge, governments provide support for research, focusing on ways to address technical challenges or the need to expand the application of existing technologies. Government investment comes in the form of grants, subsidies, and procurement contracts. Intellectual property protection is also often important for supporting the efforts of firms in emerging industries. Governments may continue these roles as an industry grows, but interventions may also be more closely targeted at both product and process improvements. The role of government in the establishment of regulations and the setting of standards is likely to become more prevalent as the risks and opportunities of industries are better understood and as competition increases.

1.4 Element Interaction and Synchronization

Our framework in Figure 1.1 is built around seven underlying elements affecting the emergence of industries. We also illuminate the interactions and synchronization of the elements, which enables industries to emerge. We have labeled three states of emergence as Concept, Validation, and Diffusion, and each column shows how the seven elements must align for an industry to continue its emergence. Synchronization is the coordination of the elements. When these elements are in sync, an industry will continue its emergence. If, on the other hand, there is not synchronization, then the industry is likely to be stalled and hindered from further emergence.

Throughout this book we will show not only how each element changes over time, but also, and equally importantly, how the elements interact with each other for synchronization, which is necessary for industries to continue to emerge. When the elements develop to specific points at the same time, synchronization occurs, which is a key dynamic that spurs the emergence of an industry.

1.5 Conclusion

This book is organized around the seven elements summarized in the industry emergence framework (Figure 1.1). The next seven chapters describe the nature of each element, how each changes over time as industries emerge, and how each interacts with the other elements. After these seven elements are described, a chapter is dedicated to explaining how the elements interact and how synchronization is necessary for industries to emerge. The book concludes with a chapter exploring the implications of emerging industries for inventors, firms, investors, scholars, and policymakers.

Exercises

- Choose an emerging industry and describe how the elements introduced in this chapter influence its emergence.
- Choose an emerging industry, assess the interaction between the elements affecting industry emergence, and analyze whether the elements are synchronized as suggested in Figure 1.1.
- Describe how inventors, firms, investors, and policymakers are likely to affect emerging industries.

Notes

1 McGahan, A. M. (2004). How industries change. *Harvard Business Review*, October.

2 Gustafsson, R., Jääskeläinen, M., Maula, M., & Uotila, J. (2015). Emergence of industries: A review and future directions. *International Journal of Management Reviews*, *18*(1), 28–50.

3 PricewaterhouseCoopers (2013). Emerging industries: Extension of the European Cluster Observatory.

4 Aldrich, H., & Fiol, M. (2007). Fools rush in? The institutional context of industry creation. In A. Cuervo et al. (Eds.), *Entrepreneurship Concepts, Theory and Perspective* (pp. 105–129). New York, NY: Springer Publishing.
5 Calori, R. (1990). Effective strategies in emerging industries. In R. Loveridge, & M. Pitt (Eds.), *The Strategic Management of Technological Innovation.* New York, NY: John Wiley.
6 Goldstein, J. (1999). Emergence as a construct: History and issues. *Emergence, 1*(1), 49–72.
7 Porter, M. (1980). *Competitive strategy: Techniques for analyzing industries and competitors.* New York, NY: Free Press.
8 Porter, M. (1980). *Competitive strategy: Techniques for analyzing industries and competitors.* New York, NY: Free Press.
9 Porter, M. (1996). What is strategy? *Harvard Business Review,* November–December, 61–78.
10 Lamoreaux, N. R., & Sokoloff, K. L. (Eds.) (2007). *Financing innovation in the United States: 1870 to the Present.* Cambridge, MA: MIT Press.
11 Hung, S., & Chu, Y. (2006). Stimulating new industries from emerging technologies: Challenges for the public sector. *Technovation, 26*(1), 104–110.
12 Rogers, E. (2003). *Diffusion of innovation.* New York, NY: Simon & Schuster International.
13 Funk, J. (2010). Complexity, critical mass and industry formation: A comparison of selected industries. *Industry and Innovation, 17*(5), 511–530.

Further Reading

Forbes, D., & Kirsch, D. (2010). The study of emerging industries: Recognizing and responding to some central problems. *Journal of Business Venturing, 26*(5), 589–602.

Grant, R. M. (2010). *Contemporary Strategy Analysis.* New York, NY: John Wiley & Sons.

Gustafsson, R., Jääskeläinen, M., Maula, M., & Uotila, J. (2015). Emergence of industries: A review and future directions. *International Journal of Management Reviews, 18*(1), 28–50.

Phaal, R., O'Sullivan, E., Routley, M., Ford, S., & Probert, D. (2011). A framework for mapping industrial emergence. *Technological Forecasting & Social Change, 78,* 217–30.

2 Firm Strategy

Learning Objectives:

- Describe the role of firm strategy for industry emergence
- See how firm strategy changes with industry emergence
- Analyze how various elements affect firm strategy and how their synchronization affects industry emergence

Key Concepts:

- Firm strategy
- Industry structure
- Strategy seekers
- Standard setters
- Shakeout leaders

2.1 Introduction

Technology, investment, supply networks, production, markets, and government affect the structure and competitiveness of industries as they emerge.[1] However, the firm, which is a business organization that operates on a for-profit basis, is usually the organizational unit that advances technology, receives investment, shapes markets, organizes supply networks and production, and competes with other firms to form an emerging industry. **Firm strategy** is how a firm positions itself in an industry and how it uses resources to support its position.[2] The first firms in a new industry are strategy seekers that envision opportunities and exhibit patience while an industry grows, followed by standard setters who compete for market share as an industry begins to emerge. As an industry continues to grow, shakeout leaders seek to benefit as the focus shifts from product to process innovation. As technology and markets develop and investment increases, firms develop corresponding strategy. Firm strategy is provisional, in the sense that firms will change strategy in order to take advantage of technological or business innovations.[3] In emerging industries, the most appropriate strategy is unlikely to be apparent, so firms may experiment with a wide range of forms. Successful firms will be those whose managers are able to learn and adjust to changing conditions. As an industry emerges, managers and entrepreneurs learn from experience and react by adopting different strategies. For example, as an industry first emerges, firms may choose to license their technology such that their revenue comes from the sale of their technology or process designs, because they lack the resources to commercialize their technology. With further emergence, firms may develop their technology into components that are used for an end-market product, because the market may not have matured enough for a standalone product. With even further industry emergence, firms may develop the end-market product themselves.[4] As an industry continues to emerge, firms may vertically integrate, outsource commodity aspects of their value chains, or choose niches in the industry value chain, as all of this maneuvering leads to 'shakeout' or industry consolidation as weaker firms disappear.

2.2 *Firm Strategy and Industry Emergence*

An industry's structure and competition affect its attractiveness to firms already in the industry as well as to new entrants and influence the strategic options firms have in the industry. These options can in turn influence the emergence of an industry. The next sections present business strategy seekers, standard setters, and industry shakeout leaders, and discuss how firm strategy affects industry emergence.

2.2.1 Strategy Seekers

At the beginning of an industry, firms are usually few in number because of technology and market immaturity, and often focus on establishing their position by gaining intellectual property protection for their innovations. There may be a lack of clarity for the direction of technology as well as markets, so the pre-revenue currency is knowledge and patentable findings. It is not uncommon for business strategy options to be unclear in the earliest days of industry emergence, as companies seek to develop value in the form of intellectual property, processes, or unique content. Many of the companies active at this stage may lack saleable assets, so decisions about long-term business strategy would be premature. The extent of market demand is uncertain, the product qualities most valued by customers have yet to be determined, and all but the most adventurous investors are likely to be reluctant to commit. Innovating

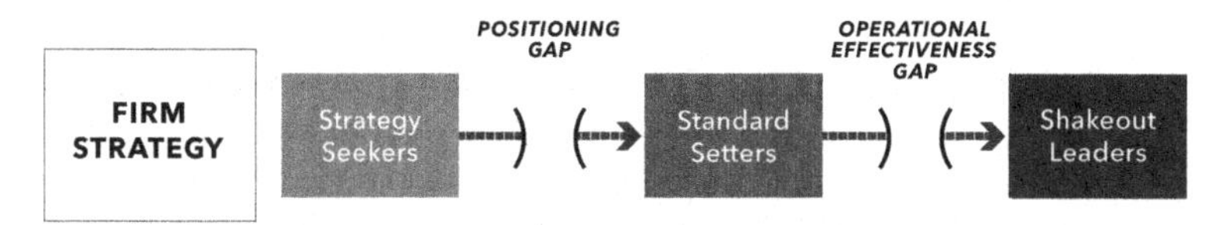

Figure 2.1 Firm Strategy and Industry Emergence.

startup firms often license or sell their technology to firms that have the substantial capital necessary to scale-up operations in order to lower per-unit costs. Firms are searching for the best business strategy for success, and ambiguity during this point of time in an industry's emergence results in a **positioning gap**, which keeps firms from setting industry standards and establishing clear strategic positioning in the industry (see Figure 2.1).[5] Research and trial and error of firm strategy options can help bridge this positioning gap.

Emerging Industry Case Study: Unmanned Aerial Vehicles

Unmanned aerial vehicles (UAV) are controlled remotely or are autonomous through software-controlled flight plans in their embedded systems, working in conjunction with the Global Positioning System (GPS). UAVs are used largely by the military, but they are also used for search and rescue, surveillance, traffic monitoring, weather monitoring, firefighting, and recreation.[6] UAV software enables remote control, flight stabilization, and the use of peripherals such as camera and video equipment. UAVs can be outfitted with gear for photography, videography, weather monitoring, weaponry, and cargo delivery.

The UAV industry has experienced significant growth as it has emerged, and companies in the industry have been seeking a strategy to position themselves as leaders in this rapidly changing industry. Industry spending was $6.4 billion in 2014 with estimates of increases to over $11 billion in the next 10 years.[7] As the industry emerges, firms are seeking profitable market opportunities and the resources to meet customer needs. For example, positioning to provide UAVs to the hobby segment of the market is challenging, due to competitive pressure from low-cost Chinese companies and uncertain regulations and consumer tastes and preferences.

(*Continued*)

The growth of the UAV industry can be seen in the growth of published documents related to the industry. According to IFI Claims, there have been 4,357 patent applications, 1,736 utility models, and 1,309 granted patents from 1994 to 2014.[8] Figure 2.2 shows the rapid growth in UAV-related publishing, especially since 2010.

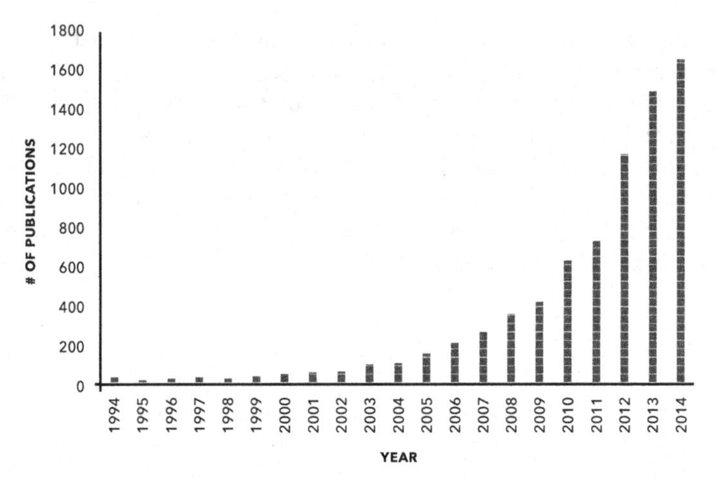

Figure 2.2 UAV-Related Publications.[9]

There are approximately 100 global UAV companies, including aerospace giants such as Boeing and Lockheed Martin and technology giants Honeywell and IBM. Smaller firms focused on non-military UAVs include 3D Robotics and DJI. The two main divisions of this industry, military and non-military, are both highly competitive due to the market power and government contract history of large firms for military UAVs and established brand, market share, and low-cost manufacturing supply chains for non-military UAV companies.

Firms in the non-military division of the UAV industry are seeking opportunities as either 'platform' companies offering products that can be adapted to many uses, or as specialty companies that focus on narrower, but deeper niche applications. These niches include entertainment and recording; journalism, filmmaking, and photography; security and monitoring; delivery and errands; and exploration, aid efforts, and disaster recovery. These opportunities all show promise, but technical capabilities and market acceptance are still developing resulting in a positioning gap, so companies in the industry are weighing their options for different positioning.

Flying UAVs has grown as a hobby and as a way to enhance other hobbies such as photography and videography. A starter UAV can cost as little as $100 and can be outfitted with a camera or video recorder. These applications offer new perspectives for filming events such as news stories, weddings, or security. In the future, UAVs may be used for errands such as delivery of prescription drugs from pharmacies, corporate documents, meals from restaurants, and food from supermarkets. Monitoring forest fires, tracking agricultural production and livestock, and inspecting oil rigs, wind turbines, and hydroelectric dams are substitute services UAVs can offer in place of endangering human lives. Options for firms have changed as the industry has emerged, revealing some industry positions to be more attractive than others, such as recreation and entertainment compared to delivery services and resulting in at least partial bridging of the positioning gap. With emergence, more types of firm positioning have appeared, and as more uses for UAVs have been developed, the industry has emerged further.

The following emerging industry case study of the UAV industry demonstrates how companies search for a business strategy in order to be competitive in an emerging industry.

2.2.2 Standard Setters

While there are usually few competitors during the early days of an industry's emergence, unless the leading patent holders license their technology widely, there are likely to be many more firms competing for the lead standard as the industry moves forward. Setting the standard for an industry is influenced by many factors, including quality and value, but also political and even social influence. In other words, the best technology does not always end up being the standard bearer, as government decree, celebrity endorsement, and alliances can influence technology and market leadership.[10] For example, companies in the digital healthcare records industry are currently competing to set the standard for the collection and use of patient health information in a digital format. The industry leader, McKesson, organized several top companies competing in the digital healthcare records industry. These companies formed the CommonWell

Health Alliance in order to establish the standard for capturing, storing, and sharing digital healthcare records. The standards these companies have developed will help their industry innovate and advance, but are also likely to help these companies maintain their industry leadership.

Companies vying for the industry standard have been able to bridge the positioning gap, which is the gap that keeps companies from differentiating themselves from an emerging industries' variety of options for meeting customer needs (see Figure 2.1). These firms are trying to establish their position in the industry and to build and acquire the resources necessary to support their position. Industry position is how a company is able to competitively distinguish itself from rivals in a way that offers customers value and enables the firm to capture value.[11]

2.2.3 Shakeout Leaders

As industries continue to emerge, there is likely to be a surge of new entrants that are product and service imitators, but that are seeking to distinguish themselves based on their expertise in process innovation. Once standards are set in an industry, competition shifts to scaling production and lowering costs.[12] Emerging industries often face an **operational effectiveness gap**, which is when companies fail to perform operational activities better than their rivals. This keeps the industry from having leading firms that can lower per-unit costs so that their industry can outcompete the current status quo providers of alternative technology. For example, the electric vehicle industry lacks clear industry leaders that have been able to scale their business in order to achieve costs that would make their products competitive with gasoline vehicles. Industries continue to emerge when a group of firms bridges the operational effectiveness gap and becomes leaders in an **industry shakeout** based on their ability to lower per-unit costs, which leads to growth in markets, investment from revenue and public offerings, and production and supply network expertise.

In many emerging industries, a shakeout results in a decline in the number of companies and an overall consolidation of the industry. Firm focus shifts to cost, pricing, and market share. Price competition and acquisitions are common signals for this state of the industry's emergence, because firms are struggling to hold market share, cover capital costs, and stay profitable, and barriers to industry entry shift from technological to capital costs for scaling and innovating production processes.

The changing industry structure leads to a different competitive environment as market demand moves from early adopters to followers,[13] with requirements, for example, for greater reliability, simplicity of operation, and lower prices. This demands a different type of technology development, the refinement of established concepts, and an emphasis on process rather than product feature improvement. The following emerging industry case study of the online video game industry demonstrates how companies are navigating business strategy options, standard setting, and industry shakeout as an industry emerges.

Emerging Industry Case Study: Online Video Games

The **online video game industry** includes video games played via the Internet. Online video games may involve large numbers of players together or solo games that are accessed via the Internet. The $54.3 billion global online video game industry can be divided into massively multiplayer online (MMO) games, mobile games, and webpage-based games, with MMO games receiving 50 percent of the users, mobile games 38 percent, and webpage games 12 percent.[14] Companies seek successful strategies by positioning in these three different segments of the online video game industry, each with different markets, for example loyal MMO gamers versus more casual mobile game users, and each with different competitive conditions, for example large dominant companies vs. many, small startups. Companies also choose different countries to position their games due to variation in tastes and preferences.

The online video game industry is made up of thousands of companies, some of which are developers (they write the game's code, design the art, and actually make the game), some of which are publishers (they make the game ready for sale by handling funding, advertising, and packaging), and some that develop and publish games. In addition to positioning in different market segments and countries, companies choose aspects of the video game value chain in which to specialize. The size of online video game companies varies significantly from large companies like Electronic Arts (EA) that has nearly 10,000 employees,[15] to many game developing companies run by one or a few individuals. These micro companies, some of which publish extremely popular games, have become easier to start and more widespread thanks to crowdfunding such as Kickstarter and incubation platforms such as Steam Greenlight. For example, the most popular non-free game on iOS, Minecraft Pocket Edition, was developed by Mojang, which employs only 51 people.[16]

(*Continued*)

The online video game industry is organized geographically by segment and market. In China, which is one of the largest growth markets, MMO and webpage-based games are most popular, and there are many developers and publishers, including Tencent QQ, which has about one-third of the market share.[17] Mobile gaming is most popular in Japan, and the industry is highly concentrated by companies such as GREE and DeNA. MMO games are also most popular in South Korea, and Nexon is one of the leading companies.[18] The US market is split between MMO games, mobile games, and webpage-based games, and companies such as Zynga and Activision Blizzard have large market shares.[19]

Competition in the online video game industry is fierce due to the range of genres, geographically dispersed appeal, and the ease of game delivery, which makes for low market entry barriers. In this environment, it is very challenging to establish industry standards because game users change their preferences, and there are few mechanisms companies can use, other than constant innovation, to keep customers loyal. While there are industry leaders by country and genre, the fickleness of customers and the relatively low cost of game development and publishing make for the steady entry of new companies and new games. For example, new entrants can release games via Google Play or the Apple App Store, and word of mouth can result in rapid distribution and growth in popularity. High levels of competition fuel the industry's growth, innovation, and emergence. While growth in the global online video game industry has been steady, and the size of the global industry is expected to surpass $113 billion by 2018, with MMO games generating $34.3 billion and mobile games $30.2 billion,[20] industry conditions make company shakeout inevitable. However, even the companies that survive and become leaders face constant pressure to keep their games popular among customers of this ever-changing industry.

The global online game industry is growing rapidly and is highly competitive, offering companies opportunities along with challenges as they try to find the best strategic position, develop the leading standards, and survive company shakeout. The prospects for quick and large returns drive innovation and technological advances, leading to faster speeds and better graphics. In addition, new users and the short interest span of many existing users have resulted in a rapid innovation cycle in the industry. Companies frequently release new versions of games and supply updates with new content and features periodically. Innovation is also driven by the integration of new and better hardware developed outside of the global online video game industry, such as new smartphones, tablet computers, microprocessors, and game consoles. Low entry barriers due to low capital requirements, available distribution platforms, and advancing technology have encouraged the emergence of the industry, as evidenced by rapid growth in users, revenue, and geographic reach. The structure and competitive conditions of the global online video game industry have also aided the emergence of the industry, and thousands of companies, distributed by size, genre, and geography, have enabled many experiments with technology, themes, markets, and business strategies.

2.3 Firm Strategy and Synchronization

Synchronization of firm strategy and the other elements covered in this book is essential for continued industry emergence. Firms that are searching for the best business strategy are doing so in part because their technology and markets are in early development. Likewise, firms at this point are usually bootstrapped, with undeveloped production and supply networks. All of the elements need to be in sync in order for an industry to continue its emergence and reach the Concept state of industry emergence (see Figure 1.1).

Firms that are setting standards are usually doing so because their technology is industry leading, they have built a market of early adopters, they have secured seed funding and government protection and/or approval, they are pilot-testing their products or services, and they are growing a supply network. These accomplishments have helped these firms bridge the positioning gap in order to gain a foothold in their industry and reach the Validation state of industry emergence.

Firms that are leaders as the industry shakes out have bridged the operational effectiveness gap by having advanced their technology, grown their markets, acquired investment from revenue and public offerings, and developed production and supply network expertise. These synchronized elements enable an industry to reach the Diffusion state of industry emergence.

2.4 Conclusion

Firm strategy is how a firm responds to the conditions and structure of its industry in order to establish its position and then maintain its position by building its resources. This chapter described the development of firm strategy from strategy seeking to standard setting to shakeout leadership.

This chapter explained and illustrated firm strategy gaps (positioning and operational effectiveness) and showed how they can slow and even stall the pace of industry emergence. This chapter also helped explain why synchronization between firm strategy and the other elements is so essential for industry emergence. Firm strategy and industry structure are influenced by technological advancement and market demand, and investment and government mechanisms affect firms' strategic options. Understanding the interaction of all of the elements helps firms develop strategy and plot moves to help synchronize the elements so an industry can emerge.

Exercises

- What are the challenges of defining industry boundaries when assessing industry structure?
- How can understanding industry structure aid in understanding how an industry is likely to emerge?
- Describe how firm strategy is affected by other elements such as technology, investment, supply networks, production, markets, and government.

Notes

1 Schumpeter, J. A. (1934). *The theory of economic development: An inquiry into profits, capital, credit, interest, and the business cycle*. Cambridge, MA: Harvard University Press.
2 Porter, M. (1980). *Competitive strategy: Techniques for analyzing industries and competitors*. New York, NY: Free Press.
3 Teece, D. J. (2010). Business models, business strategy and innovation. *Long Range Planning*, 43, 172–194.
4 Fosfuri, A. (2006). The licensing dilemma: Understanding the determinants of the rate of technology licensing. *Strategic Management Journal*, *27*(12), 1141–1158.
5 Agarwal, R., & Bayus, B. L. (2004). Creating and surviving in new industries. In J. A. C. Baum, & A. M. McGahan (Eds.), *Business strategy over the industry lifecycle* (Advances in Strategic Management), Vol. 21. Bingley: Emerald, pp. 107–130.
6 Unmanned Aircraft Systems (UAS) (PDF). Available online at Icao.int (accessed January 8, 2016).
7 Teal Group (2014). UAV Market Study.
8 Available online at www.ificlaims.com/index.php?page=news&type=view&id=lcady-s-blog%2Fhovering-over-the-drone (accessed February 2, 2016).
9 Adapted from and available online at www.ificlaims.com/index.php?page=news&type=view&id=lcady-s-blog%2Fhovering-over-the-drone (accessed February 2, 2016).

10 Funk, J., & Methe, D. (2001). Market- and committee-based mechanism in the creation and diffusion of global industry standards: The case of mobile communication. *Research Policy*, *30*(4), 589–610.
11 Teece, D. (2010). Business models, business strategy and innovation. *Long Range Planning*, *43*(203), April–June, 171–194.
12 Suarez, F. (2004). Battles for technological dominance: An integrative framework. *Research Policy*, 33, 271–286.
13 Rogers, E. (2003). *Diffusion of Innovation*. New York, NY: Simon & Schuster International.
14 Available online at www.newzoo.com/insights/us-and-china-take-half-of-113bn-games-market-in-2018/.
15 EA Annual Financial Statement, 2014.
16 Available online at https://help.mojang.com/customer/portal/articles/331367-employees.
17 Regional Differences 2012 Segments. Available online at phpapp02/95/lucix-key-facts-and-trends-on-the-ever-changing-global-games-market-17-638.jpg?cb=1350555438 (accessed May 10, 2015).
18 Social gaming overview: Too big to ignore. Available online at www.slideshare.net/marksilva/social-gaming-overview-too-big-to-ignore (accessed May 10, 2015).
19 Regional Differences 2012 Segments. Available online at phpapp02/95/lucix-key-facts-and-trends-on-the-ever-changing-global-games-market-17-638.jpg?cb=1350555438 (accessed May 10, 2015).
20 Available online at www.newzoo.com/insights/us-and-china-take-half-of-113bn-games-market-in-2018/ (accessed February 2, 2016).

Further Reading

Grant, R. M. (2010). *Contemporary Strategy Analysis*. New York, NY: John Wiley.

Mintzberg, H., Ahlstrand, B., & Lampel, J. (1998). *Strategy Safari: A Guided Tour through the Wilds of Strategic Management*. New York, NY: The Free Press.

Porter, M. (1980). *Competitive Strategy: Techniques for Analyzing Industries and Competitors*. New York, NY: Free Press.

3 Technology

Learning Objectives:

- Explain how technology is made up of design aspects, knowledge, and physical components
- Describe the processes of technology research, discovery, viability, and deployment
- Show how a technology combines with other technologies to begin the emergence of an industry
- Understand how the advancement of technology can be measured and evaluated for its effect on industry emergence

Key Concepts:

- Research and Discovery
- Viability
- Deployment
- Technology contingency
- Technology readiness

3.1 Introduction

The development of a technology or group of technologies is the seed that begins most new industries. However, while technology may be necessary for an industry to emerge, it is not individually sufficient. The other elements covered in the industry emergence framework in Chapter 1 (Figure 1.1) interact with technology in order for an industry to emerge and grow.

This chapter presents the role of technology in the emergence of new industries. **Technology**, broadly defined, is made up of design aspects, knowledge, and usually physical components. The design aspects are the needs that motivate the advancement of the technology. The knowledge includes know-how and can also be embodied in physical devices and equipment, that is, physical components.[1] A key concept underpinning how we see technology is that "while individual technologies improve, they also combine with others to make existing products more useful, and to make new ones possible."[2] This chapter presents an understanding of how technology affects industry emergence; how an industry's technology can be a collection of design aspects, know-how, and physical components; how a technology combines with other technologies to begin the emergence of an industry; and how the advancement of technology can be measured and evaluated for its effect on industry emergence.

3.2 Technology and Emerging Industries

Research and the discovery of technology are usually the start of new industries. **Research** is study directed toward greater knowledge or understanding of the fundamental aspects of a field or topic. This new knowledge is combined with other knowledge,[3] design aspects, and usually physical components in order to develop new technology. Technology is critical for industry emergence, and it is influenced by other elements. The influence of other elements moves technology forward in a synchronized fashion from research and discovery to viability to deployment (see Figure 1.1 and Figure 3.1).

Technology becomes **viable** when there is evidence that it accomplishes an intended feat and when a process occurs or a device or substance is produced with the reasonable chance of replication. Viability is a critical step for emerging industries because it signals to researchers, engineers, and scientists that the state of a technology has moved past conceptual discovery to the point of a tangible, workable process, device, or substance. The existence of **knowledge gap**, such as missing insight or pieces of complementary technology (see Figure 3.1), can limit the progress from research and discovery to viability. Knowledge gaps can be bridged through focused research on the areas with missing insight or complementary technology. When knowledge gaps are bridged and technology viability is reached, the next steps trigger. First, knowledge of viability for a process, device, or substance helps set a standard for more focused research. Second, viability focuses other elements, such as firm strategy, investment, supply networks, production, markets, and government, onto a new process or object. Both of these reactions aid in further development of the technology and with deployment of the process or object.

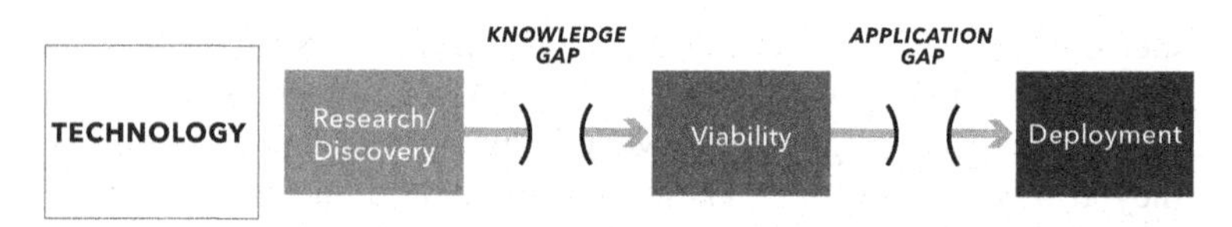

Figure 3.1 Technology and Industry Emergence.

When technology is ready for **deployment**, it is available to help address a technical challenge, and scientists and engineers are able to bridge an application gap (see Figure 3.1). An **application gap** exists when a technology is viable (useable), but does not provide a solution to a technical challenge. Searching for ways to adapt the technology in order to solve a challenge, or by finding a challenge for the technology to solve that had not been previously addressed, can bridge an application gap. Technology viability and deployment signal potential uses for the technology. Technology deployment does not necessarily require the development of a market. Instead, deployment is the advancement of the technology to the point where, if there are firms, funding, supply networks, production capabilities, demand, and government approval to supply the necessary complementary pieces of knowledge and tangible components, then a market may form for the technology.

3.3 Technology Contingency

The emergence of an industry is dependent upon the development and advancement of multiple technologies. We call this "technology contingency." **Technology contingency** is the realization that few, if any, technologies succeed independent of other complementary technologies. This is directly relevant to the study of emerging industries because industries continue to move forward only when the necessary pieces of technology fit together to support each other and form a complete technology solution. Just as a puzzle needs all of its pieces for a complete picture, emerging industries are incomplete and held back when complementary technologies are missing. This "technology hold-up" is a valuable planning concept, as it focuses attention on a group of technologies and processes instead of simply on the advancement of a sole technology. Firms and their partners can use the technology contingency concept to develop a broader and more inclusive focus to envision, plan for, and encourage industry emergence.

Technology contingency is relevant for most industries. For example, the regenerative medicine industry depends upon stem cell extraction, cultivation, and implantation technologies, each of which is at a different state of development. The electric vehicle industry relies on batteries and battery charging technology, both of which have limited the emergence of the electric vehicle industry. These two examples show the importance of assessing the development of all relevant technologies: the primary or core technology or technologies of a particular emerging industry, as well as complementary technologies, in order to focus technology development efforts and make more accurate judgments of the prospects for the emergence of industries. The following is an emerging industry case study on the wearable healthcare device industry that illustrates how the integration of multiple technologies affects industry emergence.

Emerging Industry Case Study: Wearable Healthcare Devices

The **wearable healthcare device industry** develops wearable electronics, or biosensors, that enable health monitoring in real time and allow for interpretation of individual data such as heart rate, steps taken, glucose level, and temperature. There is the potential for wearable devices to connect patients more closely to their doctors and for patients to play a more active role in their healthcare and wellbeing. Wearable devices are made up of multiple technologies such as sensors, electronics, and software, all of which must work together in order to successfully record, interpret, and communicate healthcare data. Example devices include activity trackers, smart watches and clothing, patches and tattoos, and smart implants.

The emergence and growth of the wearable healthcare device industry has been spurred by all of the seven elements introduced in Chapter 1: firm strategy, technology, investment, supply networks, production, markets, and government. Figure 3.2 shows the growth of the wearable device industry, including wearable healthcare devices over the past five years and the projections going forward. By 2015, wearable healthcare devices topped $1 billion in market size.[4]

The emergence of the wearable healthcare device industry has been driven by the development of technology over the past few decades. The number of wearable device patents filed has grown from single digits per year in the 1980s to a 40 percent compound annual growth rate between 2010 and 2015.[6,7]

During the beginning of its emergence, the wearable healthcare device industry experienced a setback as consumers became dissatisfied with the insufficient data provided by the first healthcare wearables (known as activity trackers), and preferred more

(*Continued*)

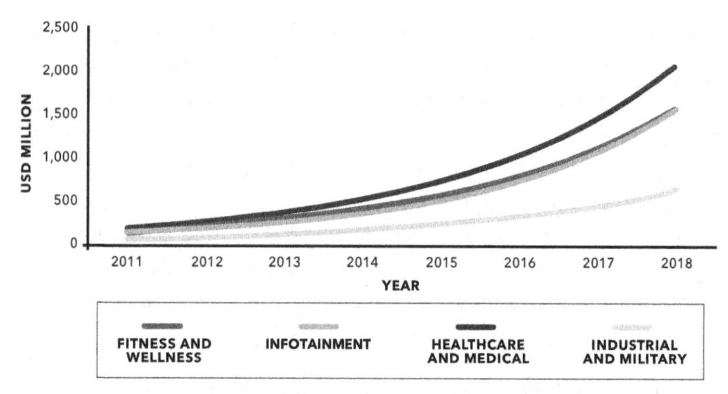

Figure 3.2 Growth of Wearable Technology.[5]

personalized and better-interpreted data. On average, users stopped using the devices after just six months.[8] While the lack of capabilities and limited data analysis cause the technology to be seen as a bottleneck for the industry, it has actually helped push the industry further towards emergence. Consumers' limited adoption of the products has motivated companies to work to address the barriers, and as a result, the industry has experienced breakthroughs in technology, which in turn has led to advances in product innovation and capabilities. Advances in firm strategy, technology, investment, supply networks, production, markets, and government needed to be in sync in order for the state of emergence of the wearable healthcare device industry to move forward from Concept to Validation and more recently on toward Diffusion.

The wearable healthcare device industry also exhibits the importance of the advancement of multiple technologies (technology contingency) in order for the industry to emerge. Sensors are critical for the industry, but so are electronics, data transfer and storage (including cloud computing), and data analytics. With advances in technology and the corresponding increases in capabilities and portability, markets have expanded to athletic, health, and military applications. These new technology-driven market opportunities have attracted a range of investments, from government grants and procurement contracts to venture capital and initial public offerings. Advances in complementary technologies such as smart phones have made wearable healthcare devices easier to use, helping connect patients more closely to their doctors and helping empower patients to monitor and improve their health. People are better able to monitor their health in ways such as keeping track of their exercise levels and controlling their diet, due to their increased ability to track their glucose levels.

Wearable healthcare devices have the potential to lead a critically important trend toward personalized, preventative healthcare. With the use of devices such as activity trackers, smart watches and clothing, patches and tattoos, and smart implants, patients have information that enables them to change their lifestyle and improve their health and wellbeing. This industry has the potential to release patients from the burden and restraints of being hospital-bound, giving them more freedom and hopefully encouraging better adherence to treatments and hence better health.

3.4 Assessing the Development of Technology

Assessing the development of technology offers insight into the pieces of an industry's technology puzzle, as well as more useful judgments of how an industry is emerging. The first step in assessing the development of technology is to separate the pieces of the industry's technology. This begins by focusing on the primary or core technologies, but it should also include any critical complementary technologies that are likely to be holding up the emergence of the industry. For example, a technology review of the wearable healthcare device industry usually would be centered on data sensor technology, but it is important to also consider data storage and transfer technology and data analytics as complementary technologies that may affect the emergence of the industry.

The next step is to evaluate the state of development of each technology. The most commonly used approach for assessing technology development is based on the US National Aeronautics and Space Administration (NASA) Technology Readiness Levels (TRL).[9] The TRL approach uses a scale of technological development running from level one (basic research principles observed and recorded) to level nine (technology proven through successful operation). The following are the nine NASA Technology Readiness Levels.[10]

- TRL 1: Scientific research has begun and results are being translated into future research and development.
- TRL 2: Basic principles have been studied and practical applications can be applied to the initial findings.
- TRL 3: Active research and design begin, generally requiring both analytical and laboratory studies in order to see if a technology is viable and ready to proceed further through the development process.

- TRL 4: Multiple component pieces are tested with one another.
- TRL 5: Technology undergoes more rigorous testing, with simulations run in environments that are as close to realistic as possible.
- TRL 6: Technology has a fully functional prototype or representational model.
- TRL 7: A working model or prototype of the technology is demonstrated in the environment envisioned for future use, for example, in space.
- TRL 8: Technology has been tested and "flight qualified" and it is ready for implementation into an already existing technology or technology system.
- TRL 9: The technology has been "flight proven" during a successful mission.

This assessment of technology development roughly corresponds to TRL 1, 2, and 3 occurring during Research and Discovery, TRL 4, 5, and 6 during Viability, and TRL 7, 8, and 9 during Deployment (as shown in Figure 3.1) and is critical to the understanding of emerging industries because it provides insight into some of the likely bottlenecks in industry emergence as well as the time and funding needed to have a working product and technology able to advance to the point of being fully functioning. This assessment of the time needed to advance the readiness of a technology entails review and insight from industry experts, as the time varies from technology to technology. It is necessary to understand both time and funding to make assessments about the state of development of the primary or core technology and complementary technologies and how each is likely to affect an industry's emergence. The following is an emerging industry case study on the regenerative medicine industry that illustrates technology readiness aspects of an emerging industry.

Emerging Industry Case Study: Regenerative Medicine

The **regenerative medicine industry** focuses on advancing the replacement, engineering, or regeneration of human cells, tissues, or organs in order to restore or establish normal human body function.[11] The industry is advancing the regeneration of damaged tissues, organs, and functions by stimulating the body's own repair mechanisms to heal previously irreparable tissues or organs, as well as growing tissues and organs in the laboratory and safely implanting them when the body cannot heal itself. The technologies that enable tissue, organ, and function regeneration are at different points of development, and assessing their development helps to improve focus on necessary advancement in order for the industry to continue to emerge and grow.

Proof of the emergence of the regenerative medicine industry is demonstrated by company and investment growth. The industry has grown from 50 companies in 1995 to over 700 as of 2016, and investment has grown from $250 million to more than $6 billion over the same period.[12,13]

Regenerative medicine technologies enable the biopsy of stem cells, the isolation and expansion of stem cell material, and the seeding, growth and regeneration of human cells, tissues, and organs. These technologies are at different levels of advancement, and their successful use depends on their application. For example, regeneration of skin and cartilage is well advanced and currently in use, while regenerated organs are still in early development, as analytical and laboratory studies are currently underway. Figure 3.3 combines several concepts from earlier in this chapter to show states of emergence, technology contingency, and technology readiness for the regenerative medicine industry. Seeding and growth of human cells are largely at Research and Discovery and TRL 1, 2, or 3, and held back by a knowledge gap. Isolation and expansion of human cells have advanced to Viability and TRL 4, 5, or 6, but are held back by an application gap, and biopsy techniques are deployable and at TRL 7, 8, or 9.

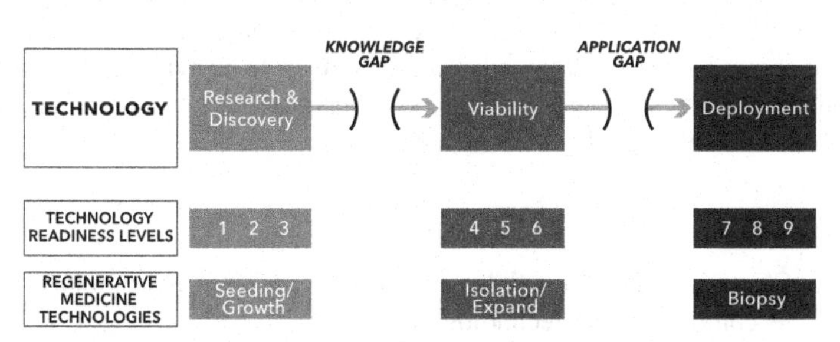

Figure 3.3 Technology and Regenerative Medicine Industry Emergence.

Markets are robust for regenerative techniques, as joint replacement and organ transplant are the current dominant technologies, with early adopters driving interest in alternatives using regeneration. Historically, government research funding has been the primary form of investment, though the past few years have seen increases in venture capital and corporate investment. The

(*Continued*)

industry's production and supply networks are at the pilot and collaborative levels respectively. Technology advancement, steady government support, and production techniques are the primary bottlenecks affecting the regenerative medicine industry's emergence. Applying the states of emergence shown in Figure 1.1, the seven elements affecting the regenerative medicine industry have synchronized enough in order for the industry to advance from the state of Concept to the state of Validation. Further individual element advancement and synchronization will be necessary for the regenerative medicine industry to continue its emergence to the state of Diffusion.

3.5 Conclusion

Technology usually plays a prominent role in the emergence of an industry. This chapter describes the advancement of technology from research and discovery to viability to deployment and links each step to industry emergence. In addition, this chapter presents technology contingency as a central concept for understanding that primary or core technology is the focus and often the driver of new industries, but that it often must be coupled with supporting technologies. If these complementary technologies are lacking, then advancement of the industry will be contingent upon the development of these support technologies. This chapter also offers an approach for assessing the influence of the degree of advancement of various technologies and their effects on industry emergence, and shows how technology advancement needs to be synchronized with other elements such as firm strategy, investment, supply networks, production, markets, and government.

Exercises

- Choose an industry and describe its primary or core technology and its complementary technologies.
- Assess the readiness of technologies using the Technology Readiness Levels that are relevant to an industry of your choice and explain how each technology's readiness affects the emergence of the industry.
- Describe how an industry's technology advancement is affected by other elements such as firm strategy, investment, supply networks, production, markets, and government.

Notes

1 Suarez, F. (2004). Battles for technological dominance: An integrative framework. *Research Policy*, 33(2), 271–286.
2 Marsh, P. (2012). *The new industrial revolution: Consumers, globalization, and the end of mass production.* New Haven, CT and London, UK: Yale University Press, p. 23.
3 Suarez, F. (2004). Battles for technological dominance: An integrative framework. *Research Policy*, 33(2), 271–286.
4 Transparency Market Research (2015). Available online at www.transparencymarketresearch.com (accessed February 2, 2016).
5 Adapted from Transparency Market Research (2015). Available online at www.transparencymarketresearch.com (accessed February 2, 2016).
6 Tech Corp International Consultants (2014). Number of wearable device patents filed.
7 Available online at www.developer.com/daily_news/wearables-patents-growing-40-per-year.html (accessed February 2, 2016).
8 Available online at www.techrepublic.com/article/wearables-have-a-dirty-little-secret-most-people-lose-interest (accessed December 18, 2015).
9 Mankins, J. C. (1995). *Technology Readiness Levels: A White Paper*. Washington, DC: NASA, Office of Space Access and Technology, Advanced Concepts Office.
10 Available online at www.nasa.gov/directorates/heo/scan/engineering/technology/txt_accordion1.html.
11 Wang, C. (2013). *A survey of current landscape in regenerative medicine*. Mizuho Industry Focus.
12 Nerem, R. (2010). Regenerative medicine: The emergence of an industry. *Interface*, 7, 771–775.
13 Alliance for Regenerative Medicine (2016). State of the industry.

Further Reading

Funk, J. (2012). *Technology Change and the Rise of New Industries*. Palo Alto, CA: Stanford University Press.

Mankins, J. C. (1995). *Technology Readiness Levels: A White Paper*. Washington, DC: NASA, Office of Space Access and Technology, Advanced Concepts Office.

Marsh, P. (2012). *The new Industrial Revolution: Consumers, Globalization, and the End of Mass Production.* New Haven, CT and London, UK: Yale University Press.

Nerem, R. (2010). Regenerative medicine: The emergence of an industry. *Interface*, *7*, 771–775.

Suarez, F. (2004). Battles for technological dominance: An integrative framework. *Research Policy*, *33*(2), 271–286.

4 Investment

Learning Objectives:

- Describe the role of investment in the birth and growth of industries
- Identify the types of investment relevant for emerging industries
- Apply the concept of investment gaps to explain the emergence of industries
- Analyze how various elements affect investment and how their synchronization affects industry emergence

Key Concepts:

- Bootstrap funding
- Seed funding
- Angels
- Venture capital
- Growth funding
- Corporate investment
- Risk–reward
- Milestones

4.1 Introduction

Investment, which is financial resources committed to support a venture, plays a critical role in the emergence of new industries because it facilitates the launching of firms, advancement of technology, refinement of production capabilities and supply networks, and development of markets. Bootstrap funding, seed funding, and growth funding help new industries emerge. Investment is probably the most talked about element of the emerging industry framework, but while it is important, the other elements are also important because, for example, failing technology or market gaps, incompetent firm strategy, or misdirected government policy cannot be corrected solely with more investment. However, for most emerging industries, and particularly technology-based industries, investment is critical for success.[1]

Industries usually begin their emergence with bootstrap funding, which is starting a company with one's own funds. The intent is usually to bootstrap the new venture to the point of sustainability or to the point where others invest seed funds in the venture. As industries emerge further, seed funding is prominent, which entails convincing others to invest and can take the form of government grants, angel funding, or venture capital. As industry emergence gains momentum, growth funding supports companies in the form of revenue, public offerings, and corporate investment and acquisitions.

This chapter presents categories of investment and explores the link between investment and different states of industry emergence. While investment plays a critical role in the emergence of new industries, we also assess the effects of other elements of industry emergence on investment.

4.2 *Investment and Industry Emergence*

Investment influences industry emergence, beginning with bootstrap funding during the early days of a new industry. Seed funding follows if an industry continues to emerge, and growth funding occurs with further emergence.

4.2.1 Bootstrap Funding

Bootstrap funding is founder investment for starting a company. Entrepreneurs often use these early funds to build a prototype or conduct a market study in order to convince prospective investors that their idea or discovery has potential. Entrepreneurs may bootstrap by mortgaging personal assets such as real estate, borrowing from friends and family, or even taking a personal bank loan or credit card advances. When money goes into the business, it does so as equity, not as a debt to be repaid to the entrepreneur.

A gap usually exists between bootstrap funding and seed funding, and this gap is often related to the risk and reward of investing in companies during the early days of an emerging industry (see Figure 4.1).

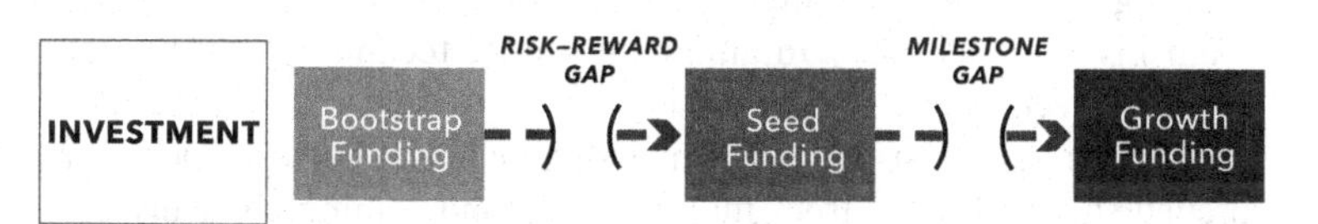

Figure 4.1 Investment and Industry Emergence.

Risk–reward assessment involves comparing the expected returns of an investment to the amount of risk needed to capture the return, and usually hinges on industry conditions. These conditions influence the accuracy of technology- and market-related projections and are influenced by economic, competitive, regulatory, and political risk. These are all context specific to the emerging industry under consideration, because they vary by industry, location, and technology. Risk–reward assessment begins with identification of the relevant economic conditions, regulations, subsidies, and tax benefits, and includes assessment of how each needs to stay unchanged or evolve for technology and market-related projections to be accurate. The time frame for variation of these circumstances should be specific to the business venture and technology, and the variations need to be included in the projections as well. Relevant insight and thorough analysis can help bridge the risk–reward gap. For example, if a pilot study reveals product functionality and market demand, this gap can be bridged and investors are more likely to provide seed funding.

4.2.2 Seed Funding

As industries emerge further, and if the risk–reward gap is bridged by sufficient technology and market progress, seed funding becomes possible. **Seed funding** entails convincing external entities to invest, and can take the form of government grants, angel funding, and venture capital.

Governments often provide seed funding in the form of research and technology grants that help the emergence of new industries. Examples include the UK Research Council funding and the US Small Business Innovation Research (SBIR) program. Many technologies need long-term, sizeable, and patient sources of funding in order to progress through research, discovery, and viability. For example, pharmaceutical companies spend $2.5 billion and 10 years, on average, to successfully bring a drug to market.[2] In comparison, successful app companies need an average of $200,000 to $1 million and one year.[3] Government funds are usually competitively awarded to universities, laboratories, and startup firms. Governments use seed funding to encourage emerging industries, projecting the benefits from investments today to anticipate future technology, firms, jobs, and tax revenues.

Angel investors, usually wealthy individuals looking to support entrepreneurship and profit from funding an up and coming company in an emerging industry, also provide seed funding. Recently, groups of angel investors have come together to try to invest more formally, sharing due diligence and contacts, structuring more formal financing terms, and involving themselves more deeply in an enterprise.

Venture capital is also a form of seed funding, and is the pooling of equity (ownership) capital that is professionally managed. General partners manage the pool in exchange for a fee and a percentage of the gain on investments. Target returns of 50 percent or higher are not uncommon hurdles for firms to apply to prospective venture capital investments.[4,5]

Risk still exists at this point for investors providing seed funding. However, most of these investors set milestones for companies to continue to receive funding. This leads to a **milestone gap** when companies fail to meet their milestones and their seed funding is retracted (see Figure 4.1). Example milestones include developing technology from a prototype to a commercial product, establishing manufacturing capabilities, and reaching sales targets. If a company meets its milestones and continues to grow with its emerging industry, the milestone gap will be bridged, and it is likely to generate growth funding.

4.2.3 Growth Funding

As industry emergence gains momentum, **growth funding** supports companies in the form of revenue, public offerings, and corporate investments and acquisitions. Firms that have a history of successful operation are likely to be able to raise public equity via an initial public offering (IPO). An IPO is the first sale of stock by a firm to the public and is usually used to grow the business through production and geographic expansion.

Corporations may play a role once the technology appears viable and the firm seems to have potential. Corporations may invest in other firms, or they may acquire the firm for its technology, market, and resources (including people). Corporate investment is increasingly seen as critical capital for bridging the milestone gap, and venture capitalists have been establishing partnerships with mature corporations in order to gain more patient investment and alliances for access to additional capital, new markets, production, and distribution.[6]

The following emerging industry case study of the social networking industry demonstrates how different types of investment play different roles in the emergence of an industry over time.

Emerging Industry Case Study: Social Networking

The **social networking industry** can be described as Internet-based software applications and interfaces that allow individuals to interact with one another, exchanging details about their lives such as biographical data, professional information, personal photos, and up-to-the-minute thoughts in order to build online communities, and includes interfaces such as chat, blogs, and sharing. The social networking concept has gone through many transformations, and what was once created for people to connect has now morphed into opportunities upon which business can capitalize. Social networking grew slowly until the mid-2000s when technology advancements helped people to be continuously connected and people's tastes and preferences shifted toward online communities. Understanding the pattern and pace of the growth of social networking reveals a story of firm strategy, technology, investment, markets, and industry dynamics, all of which influence investment in this emerging industry.

Social networking sites usually originate from social sharing ideas that need funding to advance into companies. The most likely investment available for creators of these social networking sites includes first bootstrap funding from founders, family, and friends and later seed funding from angel investors and venture capitalists. Not every type of investment is a good fit for funding a social networking startup as some are riskier, need large amounts of capital to grow, or will take longer to reach profitability. Family, friends, and angel investors often provide financial backing for small startups. The capital they provide can be a one-time injection of money or ongoing support to carry the company through the initial, challenging times. Venture capitalists have been the backbone for larger, high-growth social networking companies for the past 15 years. Venture capitalists usually offer their investment in return for a large portion of equity in the company. Most of the time, entities such as banks are not a likely option for new social networking sites. The reason for this is the unpredictability of the market. Banks need proof of a dependable revenue stream or a way to know that they will get their money back, because chances are that most social networking sites will struggle to get past the first phase of launching the site.

Once bootstrap and even seed funding have been obtained, the next hurdle that social networking sites face is creating revenue with their platforms. They can charge membership fees, but advertising is the biggest source of revenue, and advertising revenue for the US market is projected to grow from $6.1 billion in 2013 to $11.0 billion in 2017.[7]

The social networking industry has continued to evolve from a focus on connecting people to expanding commercial opportunities. Social networking enables targeted advertising and the acquisition of consumer data, making marketing more efficient and effective compared to traditional TV, radio, and print advertising. Sites like Facebook and Twitter are creating the majority of their revenue through advertising. The following descriptions of Facebook and Twitter offer insight on the role of investment in the emergence of the social networking industry.

Founded in 2004 by Mark Zuckerberg, Facebook has grown to be the biggest social networking site to date. The site was initially available only to Harvard students, but was later expanded to surrounding Ivy League schools, and soon after to all college students. Not long after that, the social networking site opened access to everyone around the world. Facebook did not get to be the social media giant that it is today without investment. Mark Zuckerberg and Eduardo Saverin bootstrapped the site during its startup. Eventually, they needed more financial support to grow their company, which they obtained from a group of angel investors, as well as some venture capitalists.

Facebook reports 1.44 billion active Facebook users in the world in 2015, of which 936 million visit daily and 1.25 million access the site from mobile devices.[8] In order to raise growth funding, Facebook sells ad space to companies that either purchase ad space or sign on for a premium advertising platform. This premium advertising platform allows companies to place ads both on their fans' computers and on their Facebook mobile apps.[9] Facebook's revenue comes primarily from text and display ads and is projected to grow from $8 billion in 2014 to over $20 billion in 2020.[10] Facebook has continued to introduce unique features that help keep their users, but it remains to be seen whether this will continue. The vast majority of the company's revenue stream comes directly from advertisement. If Facebook sees a drop in users, they would likely see a decrease in the number of companies advertising on their site.

Twitter was released to the public in 2006,[11] and is different than Facebook in that Twitter limits communication to 140 characters, giving people a way to interact with others that is quick and to the

(*Continued*)

point. After initial bootstrapping, Twitter raised over $60 million from venture capitalists. Twitter's first round of funding was between $1 million and $5 million. Its B round of funding in 2008 was for $22 million, and its C round of funding in 2009 was for $35 million from Institutional Venture Partners and Benchmark Capital, along with an undisclosed amount from other investors.[12] Twitter has grown using a series of injections of venture capital funding, which have been necessary because Twitter only recently started receiving growth funding from advertising revenue. A few powerful tweets that were first-hand reports or pictures of news events popularized the concept of average people reporting breaking news before the news stations could arrive and helped propel the company into a social networking giant.

In 2015, there were over 320 million active monthly global Twitter users.[13] With this large number of users, Twitter has followed along in Facebook's footsteps and is creating revenue using ads and membership fees for advertisers. Twitter's ad revenue is projected to top $1 billion in 2015.[14] Twitter gets most of its revenue in the United States, but its revenue from outside the US is projected to grow substantially over the next few years.

The growth of social networking has revolutionized the way people interact as well as the way they get information. Social networking is not only a place for people to connect with each other; it is a way for companies to connect with people, understand their tastes and preferences, and improve their advertising. With all of the new social networking sites that enter the market, it is difficult to tell which ones will prosper and grow. Most companies start out small and utilize bootstrap funding from founders, family, and friends, and if they are successful enough to convince others to provide seed funding, angel investors and venture capitalists help push their ideas forward. The fortunate few overcome risk–reward and milestone gaps in order to attract seed funding to help them grow large enough to warrant a sustainable stream of growth funding in the form of ad revenue. Looking at the bigger social networking companies, it is clear that they were able to capture a significant number of users, but it is not clear if this is sustainable. With low barriers to entry, it is difficult to say which social networking companies will gain traction and succeed. Innovation appears to be key for social networking companies to evolve and grow, but investment and firm strategy are also important for capturing a leading global market share.

4.3 Investment and Synchronization

Synchronization of investment and the other elements covered in this book is essential for continued industry emergence. The state of emergence we refer to as Concept is supported by bootstrap funding as well as the other elements shown in Figure 1.1. Seed funding supports the state of industry emergence referred to as Validation, and growth funding supports the state of industry emergence labeled Diffusion. While investment often helps new industries emerge, growth of the other elements can attract additional investment and investment from new sources.

Research discoveries often do not occur without initial investment from individuals, governments, and universities. However, once discoveries are made, the momentum needed to attract additional funding often appears. Research results offer hope and often generate enthusiasm for viable technology, and investors begin to envision blockbuster products, new markets, and profitable business ventures. The risk–reward gap is bridged by technology and market success leading to new sources and higher amounts of investment (seed) funding (see Figure 4.1).

Supply networks and production capabilities also need investment to develop as useful assets for emerging industries. This investment can come from production or supply industries that anticipate future demand from a new and growing industry. In addition, advancements in supply networks and production often support new industries and offer their own investment opportunities.

Governments often set policy and allocate funding for technology, firms, and industries based on their speculation about the best chances for new jobs, tax revenue, and economic development. These funds have helped emerging industries such as semiconductors, biotechnology, electric vehicles, robots, additive manufacturing, and renewable energy. All of these industries and many others were aided by government seed funding, which in turn led to the development of startup firms, advances in technology, supply networks, production, technology, and markets.

The following emerging industry case study of the biofuels industry demonstrates how industry conditions affect investment and in turn synchronization and the emergence of the industry.

Emerging Industry Case Study: Biofuels

The emergence of the **biofuels industry** is the result of a combination of firm strategy, technological innovation, investment, supply networks, production, markets, and government. The process for

(Continued)

making biofuel was invented in 1898 by Rudolph Diesel, and his peanut oil-based fuel was used to power an engine.[15] First generation biofuels are made from plants such as peanuts, sugar cane, and corn, which are usually also used as food and feed. Second-generation biofuels are produced from cellulose, hemicellulose, or lignin. Major sources of cellulose are plant fibers (cotton, hemp, flax, and jute) and wood. Third-generation fuels involve advanced chemical or biological processes using waste, algae, and cellulosic feedstock. Early funding for companies developing biofuels has primarily been bootstrapping. Technical and market breakthroughs help bridge the risk–reward gap, leading to seed funding. The milestone gap has proven to be formable due to high production scaling and distribution costs, which has primarily limited growth funding to companies focused on first generation biofuels.

The early biofuels industry featured the 1908 Ford Model T, which could burn ethanol, gasoline, or both. However, the industry's growth was hindered by competitive pressure from the petroleum industry. This increased the risk and limited investment and largely stalled further emergence until the 1970s, and highlights the role of market conditions and a risk–reward gap. However, legislation, such as the US Clean Air Act of 1970, set new, more stringent standards for pollutants, increased interest in alternatives to petroleum-based fuels, helped bridge the risk–reward gap, and drove new sources of investment. The Arab oil embargoes of 1973 and 1979 added further demand for biofuels as gasoline customers in the US and Europe waited in long lines. The 2000s have seen the most significant rise in biofuels production, from approximately 300,000 barrels of biofuels per day in 2000 to over 2 million barrels of biofuels per day in 2014.[16] These historical highlights show how industry characteristics helped bridge the risk–reward gap, drove seed and growth funding, and affected the emergence of the biofuels industry.

4.4 Conclusion

Investment enables research, innovation, and commercialization, and is important for the emergence of new industries because it funds the launch of firms, the advancement of technology, the refinement of supply networks and production capabilities, and the development of markets. This chapter described different categories of investment (bootstrap funding, seed funding, and growth funding) and when and why they occur, and it explained and illustrated investment gaps (risk–reward and milestone) and showed how these gaps can slow and even stall the pace of industry emergence. This chapter also emphasized the importance of multiple elements and their synchronization for industries to emerge. Investment is critical for the birth and growth of new industries, and it is in turn dependent upon technology development and market demand. Likewise, as shown in the social networking and biofuels case studies, investment is often also dependent upon elements such as firm strategy, government mechanisms, production advancements, and the support of supply networks.

Exercises

- Choose an industry and describe how different types of investment have played a role as the industry has emerged.
- Compare the motivation of providers of different types of investment.
- Describe how investment is affected by other elements such as firm strategy, technology, supply networks, production, markets, and government.

Notes

1 Lamoreaux, N. R., & Sokoloff, K. L. (Eds.) (2007). *Financing Innovation in the United States: 1870 to the Present*. Cambridge, MA: MIT Press.
2 Available online at www.scientificamerican.com/article/cost-to-develop-new-pharmaceutical-drug-now-exceeds-2-5b/.
3 Available online at http://thenextweb.com/dd/2013/12/02/much-cost-build-worlds-hottest-startups/.
4 Kenney, M. (2010). Venture capital investment in greentech industries: A provocative essay. In *Handbook of Research on Energy Entrepreneurship*. New York, NY: Edward Elgar.
5 Dimov, D., de Holan, P. M., & Milanov, H. (2012). Learning patterns in venture capital investing in new industries. *Industrial and Corporate Change*, 21, 1389–1426.
6 Dittmer, J., McCahery, J., & Vermeulen, E. (2014). The "new" venture capital cycle and the role of governments: The emergence of collaborative funding models and platforms. In *Boosting open innovation and knowledge transfer in the European Union*. Directorate-General for Research and Innovation. Brussels, BE: European Commission.
7 BIA Kelsey. Available online at www.biakelsey.com.
8 Available online at http://venturebeat.com/2015/04/22/facebook-passes-1-44b-monthly-active-users-1-25b-mobile-users-and-936-million-daily-users/.
9 Levin, L. (2013). Available online at http://lauraleewalker.com/2012/03/28/4-ways-social-media-sites-make-money
10 Rae, S. (2013). Bubbles, big data and the future of the social media industry. Trefis. Available online at www.trefis.com/stock/idea/articles/206600/bubbles-big-data-and-the-future-of-the-social-media-industry/2013-09-19.

11 Wire, C. (2013). Looking at Twitter's history as it goes public. ABC 15. Available online at www.abc15.com/dpp/news/national/looking-at-twitters-history-as-it-goes-public.
12 Jana, D. (2013). Inspiration, laboratory. Useful information about the history of Twitter and how it starts, works, criticized and many more. Available online at http://djdesignerlab.com/2009/11/01/useful-information-about-history-of-twitter-and-how-it-starts-works-criticized-and-many-more/.
13 Available online at www.twitter.com.
14 Richter, F. (2013). Twitter ad revenue tipped to double in 2013. Available online at www.statista.com.
15 Schmidt, C. (2007). Biodiesel: Cultivating alternative fuels. *Environmental Health Perspective*, *115*(2): 86–91.
16 International Energy Administration (2015). Energy statistics. Available online at www.iea.org.

Further Reading

Kenney, M. (2010). Venture capital investment in greentech industries: A provocative essay. *Handbook of Research on Energy Entrepreneurship*. New York, NY: Edward Elgar.

Lamoreaux, N. R., & Sokoloff, K. L. (Eds.) (2007). *Financing Innovation in the United States: 1870 to the Present*. Cambridge, MA: MIT Press.

Stevenson, H., & Roberts, M. (2002). *New Venture Financing*. Boston, MA: Harvard University Press.

5 Supply Networks

Learning Objectives:

- Learn how supply networks support the emergence of new industries
- See how supply networks evolve from legacy to collaborating to lead suppliers as an industry emerges
- Understand how firm strategy, technology, investment, production, markets, and government affect supply networks

Key Concepts:

- Supply networks
- Legacy suppliers
- Collaborating suppliers
- Lead suppliers

5.1 Introduction

New industries are seldom formed around a single technology or by a single entity. Instead, multiple technologies, inputs, partners, and alliances are developed in new ways to meet new needs. For example, in the digital health industry, researchers and firms offer new ways to improve health with diagnostic tests and wearable devices by combining advances from suppliers in industries such as biotechnology, electronics, and software. **Supply networks**, which we describe as groups of individuals, firms, and institutions providing inputs for the development of a product or service, form and grow as an industry emerges. Supply networks are usually made up of legacy suppliers when the industry they support is beginning. Most emerging industries are based on new technology, and therefore, related supply networks are usually based on existing industries that need to innovate and grow as well. As an industry continues to emerge, it does so in part because related and supporting entities collaborate to develop solutions. As a new industry emerges enough to begin to diffuse, its supply network begins to experience its own shakeout and consolidate around leading suppliers with technological and/or market dominant positions.

Supply networks are not independent of the other elements in the emerging industries framework. Supply networks depend on investment in technologies and processes, which is influenced by market growth, firm strategy, and the development of production. This chapter presents the role of supply networks in support of emerging industries, how supply networks progress from legacy suppliers to collaborators to lead suppliers as an industry emerges, and examines the effects of other elements such as firm strategy, technology, investment, production, markets, and government.

5.2 Supply Networks and Industry Emergence

Supply networks help develop knowledge, technology, and business models to drive industry emergence.[1] Innovation and technology commercialization play critical roles in the emergence of industries, and this is usually facilitated by multiple entities and their interactions.[2] Understanding supply networks and the alliances they develop can help with understanding how industries emerge. Supply network entities can work together formally and informally to bridge gaps, enabling industry emergence. The role of various entities in the birth and growth of industries helps to explain the limits, opportunities, and outcomes for the industry.

Suppliers may offer physical inputs, and they may also offer knowledge, process, and service inputs. In many cases, suppliers work collaboratively to help advance existing products or services, and they assist with moving technology forward for new products and processes. In order to understand a particular industry's emergence, it is necessary to know and understand its contributing suppliers. In addition, suppliers play important roles for shaping expectations and providing feedback for product and process improvement. Suppliers can offer co-development, piloting, and investment for advancing technology. Related and supporting industries can also provide pieces of knowledge and tangible components that have been developed outside of an emerging industry but that are essential for its emergence. Few, if any, emerging industries succeed independent of other complementary industries and their technologies.

5.2.1 Legacy Supply Networks

Supply networks change as an industry emerges. Figure 5.1 shows the progression from sparse legacy suppliers, to collaborating suppliers, to lead suppliers as an industry emerges. Early in the life of an industry, supply networks are scarce and based on **legacy suppliers**, which we define as inherited suppliers that new industries build upon as they emerge. This is necessary because the technology and processes are usually new, and few suppliers, if any, can offer needed technologies, components, and processes. Often, an industry's founding firms are forced to develop the technologies, components, and processes internally. This can have the advantage of integrating research, development, and production, aiding innovation and industry emergence. However, internal development can also be slow and can fail due to firms lacking the necessary capabilities to support their core technologies. This can result in an **expertise gap**, which is the lack of necessary technical or commercial insight from suppliers, and can stall industry emergence. This gap can be bridged by the collaborative efforts and advances made by an emerging industry's supply network. For example, whole-organ transplantation, such as the first heart transplant in 1967, could not occur until there were machines to sustain life during the operation, tools for the surgeons to operate and to repair the wounds they created, and methods for preserving organs during transport.[3]

5.2.2 Collaborative Supply Networks

As an industry emerges and more firms join in, the supply network also grows to meet the new demand. There is often a disaggregation of the industry's product into components made by **collaborating suppliers** as they develop specialties, grow their expertise and volume, and take advantage of economies of scale. This process parallels the industry's emergence from initial very sparse supply networks to dense subsectors of expertise to meet the various needs of the new and growing industry.

The relationships between different supply network entities affecting industry emergence are far from random, but not by any means scripted. The alliances that bridge gaps in the industry emergence process evolve and are shaped by external conditions and the relationships between entities. Firms in emerging industries are able to innovate, grow, and survive if they can acquire critical resources from external entities that help

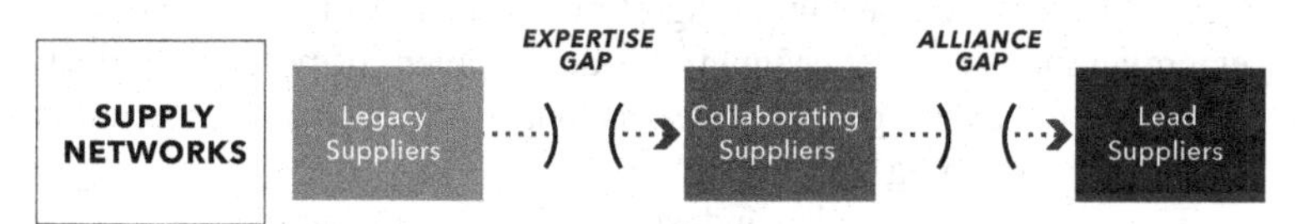

Figure 5.1 Supply Networks and Industry Emergence.

them extend their boundaries and compensate for their shortcomings.[4] Emerging industries need suppliers to develop alliance capabilities in order to overcome the gaps in the progression from an initial idea to an established industry. An **alliance gap** develops if suppliers cannot help firms in emerging industries exploit new technologies with minimal internal resources. This gap is bridged if suppliers can increase access to technology or to market expertise, reduce risk and financial exposure, and improve focus when bringing resources to bear on innovation.[5] For example, the endorsement implied by alliances between firms with new disruptive technology and established entities (other firms and even government agencies) can signal the viability of new technology and products, thereby establishing legitimacy for the new industry.[6] The networks and connectivity among industry entities enable bridging of the alliance gap in the emergence process of an industry. These relationships allow for the utilization of external knowledge, capital, and political influence to help address the pitfalls and deterrents to technology advancement, firm growth, and industry birth and emergence.

5.2.3 Lead Suppliers

Supply networks are likely to concentrate as more successful, **lead suppliers** acquire smaller, less healthy firms as an industry continues to emerge. This consolidation helps new industries emerge even further and faster, because the consolidating lead suppliers focus on process innovation and better consistency of technologies, components, or processes, and are able to lower their costs and their prices.

Lead suppliers emphasize their core resources and capabilities and focus on improving their efficiency and profitability.[7] Their consolidation targets are underperformers and innovative startup competitors. Slower firms that do not excel at process innovation eventually become acquisition targets and are likely to disappear. Some suppliers may compete on quality to separate their offerings from lower-cost products or services, while other suppliers may try to compete to have lower costs and prices by increasing their volume of sales and making profits from greater inventory turnover. The surviving lead suppliers will be larger and more dominant companies. Supplier consolidation can help emerging industries because the strongest suppliers can offer companies in emerging industries better inputs, co-development of products and services, and lower component prices, as long as there is some competition among the remaining suppliers.

The following emerging industry case study of the electric vehicle industry shows how supply networks grow from scarce legacy suppliers to collaborating suppliers to lead suppliers as an industry emerges.

Emerging Industry Case Study: Electric Vehicles

The **electric vehicle industry** entails the development, production, and maintenance of vehicles with a rechargeable battery that operate using an electric motor instead of an internal combustion engine. Evidence of this industry's emergence is that there were fewer than 1,000 electric vehicles on the roads in 2010,[8] but by 2015, there were over 1,000,000.[9]

The electric vehicle industry has been emerging due to technological innovation, government investment and incentives, venture capital, market acceptance, and the growth and strengthening of supply networks. The core technologies for an electric vehicle, which are the electric motor and the battery, have been available and in use for over 100 years, though electric vehicles were largely ignored, at least for widespread application, when the internal combustion engine became the industry standard in the early 1900s.[10] Legacy suppliers of electric motors and batteries existed and provided the core inputs to the electric vehicle industry. However, limited market demand reduced the availability of investment and the commercial incentive to adapt these technologies further. In other words, the firm strategy, technology, investment, supply networks, production, markets, and government intervention were not adequately synchronized in order for the electric vehicle industry to continue its emergence.

While the electric vehicle industry has grown rapidly since 2010, annual global sales still represent less than 1 percent of total annual global sales of all vehicles.[11,12] The constraints, or bottlenecks, experienced during the recent emergence of the electric vehicle have been related to technology, markets, investments, and supply networks. Many of the critical challenges begin with supplier issues related to technologies complementary to electric motors, such as those associated with batteries, power electronics, drive train components, and charging stations. For example, an expertise gap exists in that current technologies result in batteries that have proven to be expensive, to generate too much heat, and to offer too few hours of dependable charge. The result has been consumers that have proven fickle about adopting electric vehicles as gasoline prices have fluctuated. It appears that price sensitivity is more powerful for this industry than concerns about climate change, local pollution, and dependence on foreign oil. There has been a group of early adopters, but by and large, it has taken consumer tax credits and price shocks to sway these buyers. Suppliers will need to advance battery and

(*Continued*)

charging station technology even more in order to bridge the expertise gap the electric vehicle industry currently faces. Figure 5.2 shows the progress Li-Ion battery suppliers have made at lowering US$/Wh, showing how suppliers can affect the price of a critical input component and hence the emergence of the industry.[13]

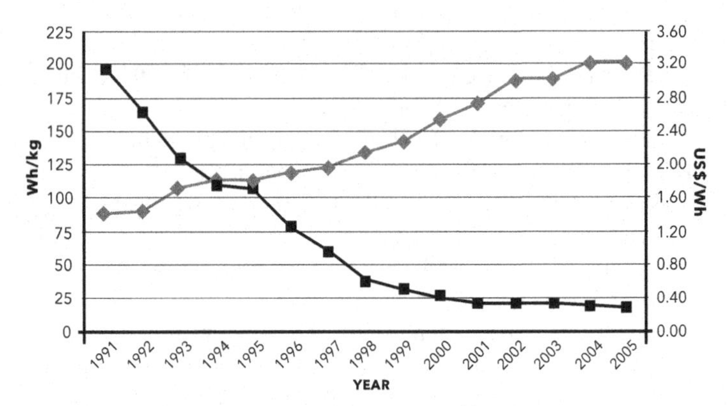

Figure 5.2 Li-Ion Pricing and Energy Density.[14]

Nissan and Tesla have been the early leaders to introduce electric vehicles. Nissan has sold more than 170,000 of its Leaf electric vehicles since 2010,[15] and Tesla, a Silicon Valley startup, which has received significant government and venture funding and launched a public offering, has sold close to 100,000 of its expensive luxury sedans. Both companies are testing the feasibility and technological readiness of the industry. However, established automakers such as Toyota, GM, and Ford have introduced or are planning to introduce electric vehicles, and it would be naïve to overlook the long-term influence of these carmakers in an industry controlled by a small number of large companies.

Multiple factors, including industry size and structure, can affect an industry's emergence. In the electric vehicle industry there are at present both young firms funded by venture capital (Tesla) and major corporations (Toyota, GM, Ford, Nissan) able to rely upon their own internal resources, but compared to many industries, there are few competitors because high capital and learning costs have slowed technological innovation and product value and limited market demand. In addition, legacy suppliers have helped the electric vehicle industry so far, but real advances for the industry are dependent upon the bridging of the expertise and alliance gaps and collaborative suppliers working with electric vehicle companies to substantially advance batteries, power electronics, and charging stations.

5.3 Supply Networks and Synchronization

Firm strategy, technology, investment, production, markets, and government can influence advances in supply networks. Improvements in technology and increased demand for materials and processes can open new possibilities for supplying materials and parts to an emerging industry. Investment in suppliers and new supplier strategies can benefit an emerging industry. Government investment and promotion and the production improvements made by related and supporting industries can help advance an emerging industry. Therefore, all of the other elements discussed in this book can affect supply networks, and together they can influence the emergence of new industries. The following emerging industry case study of the wind turbine industry demonstrates how supply networks affect the emergence of an industry.

Emerging Industry Case Study: Wind Turbines

The global **wind turbine industry** began a period of significant emergence in 1990 and has grown rapidly over the past 25 years (see Figure 5.3). The industry has attracted over $5 trillion in investment, supports more than 1 million jobs, and currently contributes 5% of total global installed electricity generation capacity.[16] Installed capacity grew to nearly 370 gigawatts (GW) in 2014, led by the US, Germany, Spain, China, and India.[17] Strong growth supported by government, social, and economic drivers is expected to continue. Legacy suppliers have determined the agglomerations of wind turbine manufacturers and installations.[18] Collaborating and lead suppliers have helped the wind turbine industry move forward with better materials, components, and equipment.

Supply networks made up of firms in complementary or support industries such as iron foundries, metal fabrication, motors and generators, electronic equipment, and power transmission have aided the emergence of the wind turbine industry. Advances in these industries, especially in close collaboration with wind turbine companies, have improved the efficiency and durability of wind turbines, enabling wind power to be a least cost alternative in some parts of the world.[20] Wind turbine manufacturers benefit from collaborative alliances with their suppliers. These relationships help wind turbine manufacturers understand their options when they are

(*Continued*)

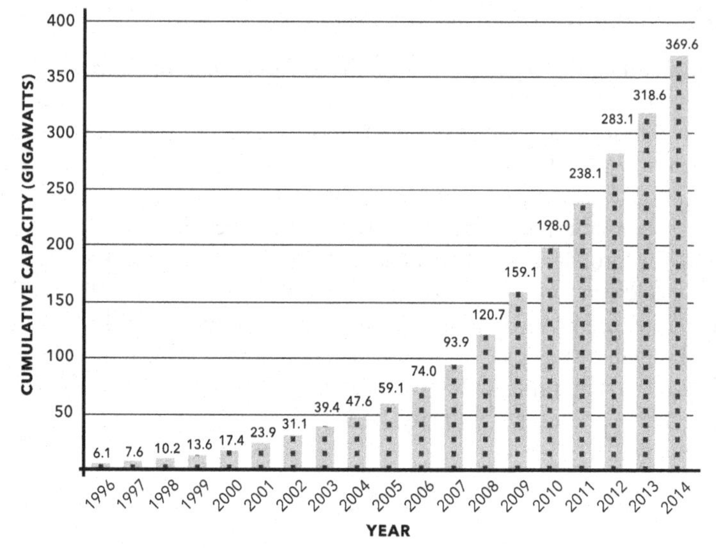

Figure 5.3 Global Wind Power Capacity.[19]

addressing a problem or developing new technology. These suppliers are a source of knowledge or can help develop new knowledge by promoting the exchange of information and the pooling of resources to jointly design new components and products. Legacy, collaborating, and lead suppliers have played important roles in supporting the wind turbine industry's emergence.

5.4 Conclusion

Supply networks are often a critical element for the formation and growth of new industries. These networks offer legacy technology and knowledge as new industries first emerge, they specialize as they collaborate for product and service improvements as an industry continues its emergence, and they later become consolidated groups of lead suppliers focused on efficiency and process improvements that help an emerging industry grow and diffuse. The strongest supply networks encompass a wide variety of support industries, which aids in the development of alliances and benefits the emerging industry, facilitating innovation and the commercialization of technology. Supply networks are also dependent on other elements, and synchronization between the elements covered in this book is necessary for industries to continue to emerge. Technology improvements and increased demand for materials and processes can open new possibilities for supplying materials and parts to an emerging industry. Investment in suppliers and new supplier strategies can benefit an emerging industry. Government investment and promotion and the production improvements made by related and supporting industries can also help advance an emerging industry.

Exercises

- Choose an industry and describe the entities that make up its supply networks.
- Describe how supply networks can change during the emergence of an industry from legacy, to collaborating, to lead suppliers.
- Explain how firm strategy, technology, investment, production, markets, and government affect supply networks.

Notes

1 Adner, R., & Levinthal, D. (2002). The emergence of emerging technologies. *California Management Review*, *45*(1), 50–66.
2 Forbes, D., & Kirsch, D. (2010). The study of emerging industries: Recognizing and responding to some central problems. *Journal of Business Venturing*, *26*(5), 589–602.
3 Saltzman, M. (2015). *Biomedical Engineering: Bridging Medicine and Technology*. Cambridge, UK: University of Cambridge Press.
4 Dyer, J. H., & Singh, H. (1998). The relational view: Co-operative strategy and sources of inter-organizational competitive advantage. *Academy of Management Review*, *23*(4), 660–679.
5 Hess, A. M., & Rothaermel, F. T. (2011). "When are assets complementary? Star scientists, strategic alliances, and innovation in the pharmaceutical industry. *Strategic Management Journal*, *32*(8), 895–909.
6 Arikan, A. M., & McGahan, A. M. (2010). The development of new capabilities in firms. *Strategic Management Journal*, *31*(1), 1–18.
7 Deans, G., Kroeger, F., & Zeisel, S. (2002). The consolidation curve. *Harvard Business Review*, *80*(December), 2–3.
8 Available online at http://energy.gov/downloads/electric-and-hybrid-electric-vehicle-sales-december-2010-june-2013.
9 Available online at http://cleantechnica.com/2015/08/08/1-million-electric-cars-will-be-on-the-road-in-september/.
10 Ofek, E., & Ribatt, P. (2010). *Plugging in the consumer: The adoption of electrically powered vehicles in the US*. Boston, MA: Harvard Business School.
11 Available online at www.statista.com/statistics/200002/international-car-sales-since-1990/.
12 Available online at http://electricdrive.org/index.php?ht=d/sp/i/20952/pid/20952.
13 Farrell, J. (2012). Are the batteries ready? 100% clean energy requires progress on storage. Available online at http://grist.org/article/

are-the-batteries-ready-100-clean-energy-requires-progress-on-storage/ (accessed February 2, 2016).

14 Adapted from Farrell, J. (2012). Are the batteries ready? 100% clean energy requires progress on storage. Available online at http://grist.org/article/are-the-batteries-ready-100-clean-energy-requires-progress-on-storage/ (accessed February 2, 2016).

15 Available online at http://evobsession.com/nissan-leaf-sales-climb-to-over-170000-worldwide/.

16 Available online at www.gwec.net/faqs/wind-power-generates-millions-jobs-worldwide/.

17 The World Wind Energy Association (2014). *2014 Half-year Report*, 1–8.

18 Theyel, G. (2012). Spatial processes of clean technology industry emergence. *European Planning Studies*, *20*(5), 857–870.

19 Adapted from and available online at http://www.gwec.net/wp-content/uploads/vip/GWEC-Global-Wind-2015-Report_April-2016_22_04.pdf.

20 American Wind Energy Association. Available online at www.awea.org (accessed December 27, 2015).

Further Reading

Adner, R., & Levinthal, D. (2002). The emergence of emerging technologies. *California Management Review*, *45*(1), 50–66.

Deans, G., Kroeger, F., & Zeisel, S. (2002). The consolidation curve. *Harvard Business Review*, *80*(December), 2–3.

Jacobs, J. (1984). *Cities and the Wealth of Nations*. New York, NY: Random House.

Malerba, F. (2007). Innovation and the dynamics and evolution of industries: Progress and challenges. *International Journal of Industrial Organization*, *25*, 675–699.

Russo, M. V. (2003). The emergence of sustainable industries: Building on natural capital. *Strategic Management Journal*, *24*, 317–331.

Theyel, G. (2012). Spatial processes of clean technology industry emergence. *European Planning Studies*, *20*(5), 857–870.

6 Production

Learning Objectives:

- Distinguish between prototyping, piloting, and scaling production
- Learn how new methods of production make new products possible and can spur the birth of new industries
- Assess the effects of other elements such as firm strategy, technology, investment, supply networks, markets, and government on production

Key Concepts:

- Prototyping
- Piloting
- Scaling

6.1 Introduction

Production plays an essential role for industry emergence, because technology remains a curiosity if we lack the ability to turn it into a product. This chapter describes how production influences the emergence of new industries. Production can enable the conversion of an idea or invention into a product, which can lead to the birth and growth of a new industry, and new production technology can become new industries. This chapter first explains prototyping, piloting, and scaling production and then assesses the effects of other elements such as firm strategy, technology, investment, supply networks, markets, and government on production.

6.2 Production for Industry Emergence

Some people use the terms "manufacturing" and "production" interchangeably; while others draw a distinction between these two terms such that manufacturing refers to the process of turning an idea into a delivered product, and production is the actual making of the product. For the purpose of our discussion, we use the two as the same and define **production** as involving the translation of ideas into reality and often facilitating the progression from invention to a commercialized product. In the earliest stages of an emerging industry a prototype is often an important milestone. A **prototype** is a model used to test a concept. **Pilot production** is next, and consists of the testing of the production process for the quality of the output and the efficiency of the chosen processes. Once the design has been shown to be feasible, there is a wide range of production processes available depending on the quality parameters and volume required for the product. When the production processes have been established, there may be opportunities for scaling production. **Scaling production** is the focus on increasing the quantity of a good produced in order to lower the per-unit costs, because fixed costs are shared over a larger number of products produced. The scale of production is determined by combining understanding of production potential (the limits of the production techniques available) and the potential demand for production output.

The nature of production often changes with the emergence of an industry. Figure 6.1 summarizes the progression of production with the emergence of an industry. Prototyping is likely to be the first form of production as an industry emerges, as limited experience with the technology means that experimentation will be essential. As the industry continues to emerge, pilot production is likely to be more prevalent as firms attempt to replicate their prototype results and test the operability and marketability of their output. Finally, achieving economies of scale from production volume becomes a more relevant concern as the industry continues to emerge. Scaling production can take the form of batch or mass production. Batch production is when a product is created stage by stage over a series of workstations and different groups (or batches) of products are made. Mass production is when large amounts of standardized products are made in continuous fashion and often on assembly lines.

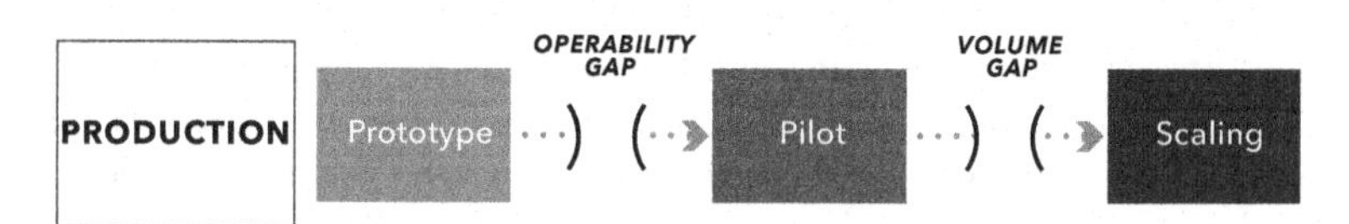

Figure 6.1 Production and Industry Emergence.

6.2.1 Prototype Production

Prototype production is often critical for moving from new inventions or ideas to products, a process often referred to as innovation.[1] Prototyping is usually the initial form of production, as an individual or firm attempts to transform an idea into a product. There are often many iterations of the prototype, as learning by doing is central to the process of turning an idea into a product. Charles Babbage (1832) explains, "we shall find in the history of each article … a series of failures, which have gradually led the way to excellence."[2] This experimentation is critical for demonstrating that an idea or invention can be turned into a product, even if only at the micro scale. A firm can gain value from prototyping production by showing that a product can be made, which can encourage further technology development, entice investment, motivate supply network development, and attract market demand.

New product ideas, and technological innovation in general, can remain underdeveloped if production methods do not exist to turn a scientific or engineering breakthrough into a product. Novel design and manufacturing technologies exemplify how production can act as a facilitator for the birth and growth of a new industry and may be necessary to enable new industries to emerge. Some of these technologies include computer-aided design, rapid prototyping, micro-fabrication, advances in materials science and nanotechnology, and novel geometries and biomimetic designs. For example, advances in rapid prototyping enable researchers and product designers to test concepts virtually or obtain physical parts in hours instead of weeks. This shortens the time to get feedback from trial runs, enabling the designer to shorten the supply chain, keeping research, design, and testing self-contained. This can also reduce errors in intent and interpretation between the steps, while enabling quicker detection of errors and less expensive design revisions. Virtual prototyping is computer-based with tests and suggestions for improvements, and additive manufacturing allows for layer-upon-layer part making to produce physical prototypes. Advances in prototyping can speed learning in industries and can facilitate the birth and growth of new industries.

6.2.2 Pilot Production

Pilot production offers proof of production capabilities for firms and can help reduce the risks of full-scale production. Production may encounter an **operability gap** if it proves difficult to replicate the prototype production (see Figure 6.1). Bridging this gap can entail improving production processes or advancing materials. If this gap is bridged, pilot production will entail smaller scale production setups that are used to generate information about the behavior of the setup for use in the design of larger production facilities. A firm may use pilot production to allow it to further develop its intellectual property and demonstrate the value of its innovations. Pilot production can allow for learning with a firm's industrial partners and further development of innovation capabilities, as well as reduce risk by identifying problems at a much smaller scale and before investing more for a larger full-scale production facility. Successful pilot production can lure new firms to the industry, further entice investment, motivate advances by supply networks, and attract market demand. In addition, pilot production can enable a company to validate its product or service with customers, which will help it make modifications to fit customers' needs. Customers also often first want to validate the functionality and useability of a product or service before signing a long-term agreement or investing in a large project. Thus, just as do the companies offering products or services, many customers prefer pilots to give them the opportunity to a) test the product or service, often referred to as proof of concept, to see how well it can work, b) develop relationships and deliver on promised outcomes, c) consider additional features that should be added to the original product or service to optimize the solution for the customer, and d) conduct a cost and value assessment of a particular product or service over the short and long term.

6.2.3 Scaling Production

Scaling production entails increasing output in order to meet market demand and benefit from economies of scale. During this process, firms may encounter a **volume gap**, keeping them from reducing cost with their increase in scale. Some processes have scale limits, and if this is the case, industry emergence can be slowed due to per-unit costs that are too high to allow an industry to be competitive with incumbent technologies and industries. If companies can find ways to reduce costs by expanding the scale of production, then the volume gap can be bridged, and the predominant expertise required for competitive advantage will begin to shift from product technology to process or production capability. Marsh (2012) describes how "as a result of time spent doing something, technical prowess [is] more or less guaranteed to improve. Along the way costs fall, while quality rises."[3] Firms' and industries' continued growth often relies on the availability of sufficient production capacity to meet demand. Cost and profit projections hinge on being able to produce in greater volumes and with higher quality. A production scaling decision is linked to customers' interest in scaled adoption, which depends on how customers assess a company's ability to offer a competitive product in terms of product functionality, useability, and cost competitiveness, as well as a company's leadership ability to develop good relationships with existing and new customers. Production scaling helps with growth readiness and product supply, but there needs to be customer demand in place to justify production scaling. A company needs to grow its production capacity based on its plans to sell the product or service.

The following emerging industry case study of the additive manufacturing industry demonstrates how production technologies can transition from prototyping to piloting to scaling for the birth of a new industry.

Emerging Industry Case Study: Additive Manufacturing

Additive manufacturing is the process of joining materials to make objects from 3D model data, usually layer upon layer. The additive manufacturing industry includes 3D printers, scanners for scanning 3D images, software for manipulating 3D images, materials ranging from plastics to metals to stem cells, and scaffolding for supporting printed objects.

Additive manufacturing is a production method that can help advance new industries as an enabling technology or tool. For example, additive manufacturing is driving disruptive biomedical innovation as it can be used to create tissue scaffolds and lattices, generate replacement tissues and potentially, in the future, build up complete organs from a patient's own stem cells. With advances in 3D printing, the regenerative medicine industry continues to emerge, and profound biomedical innovations can be put to use for improving human health.[4]

Adding material during manufacturing instead of subtracting it is revolutionary, but so is the ability to use additive manufacturing technology, for example 3D printers, to make one product at a time, on demand, in almost any location, including in space, such as in NASA's 3D printer program. This offers the potential to truly revolutionize many industries beyond just the additive manufacturing industry. The additive manufacturing industry began as prototype machines in the 1970s that made simple objects out of plastic. Pilot testing followed and improvements were made to increase quality and the range of materials that could be printed. Currently a growing number of companies sell 3D printers for business applications and over 200 companies are developing and selling 3D printers for home use.[5] The additive manufacturing industry is currently scaling production in order to lower per-unit costs. The price for consumer 3D printers has fallen to a few hundred dollars, but this still seems too high for mainstream customers. Therefore, the

(*Continued*)

additive manufacturing industry is hindered by a volume gap (see Figure 6.1), and efforts to advance all of the seven elements are underway, focusing on synchronization, which will lead to better quality, a wider range of materials, and lower per-unit costs.

6.3 Production and Synchronization

Firm strategy, technology, investment, supply networks, markets, and government can influence advances in production. Firm strategy focused on production as a competitive advantage can advance production. Technology improvements in materials and processes can open new possibilities for how to make products. Government investment and promotion and the improvements made by related and supporting industries can help advance production. All of the other elements discussed in this book affect production, and together they can influence the emergence of industries.

Another way production can influence the birth and growth of new industries is when new methods of production synchronize with the other elements and become new industries. For example, industrial robots have become an emerging industry in their own right. New production techniques may begin as a necessary advancement in order to make a new product. However, the momentum from this step can lead to whole new industries. Robotic automation, for example, began as a way to reduce labor or harm to humans during a production step, but has grown to be the foundation of production systems in industries such as automobiles and pharmaceuticals. New production technologies open new opportunities for turning ideas into products. For example, additive manufacturing began as a way to shape figures for prototyping and has expanded to include printing personalized medical devices, human tissue, entire cars and houses, and elaborate art pieces. As such, production technologies can advance and offer new applications, which in turn can become new industries. For this to happen, though, production and the other elements in this book need to synchronize. The following emerging industry case study of the nanotechnology industry demonstrates how production technologies can spur the birth of new industries through new methods of production and how these production technologies can then grow into an emerging industry. The case highlights how synchronization of the elements covered in this book enable an industry to emerge and grow from concept to validation to diffusion.

Emerging Industry Case Study: Nanotechnology

Nanotechnology is an industry based on techniques for making new materials and new products by manipulating molecules at the nanoscale (1 to 100 nanometers). Nanotechnology involves developing molecular-level systems and has affected many products such as drugs, surgical tools, solar cells, and coatings for clothing. Nanotechnology is applied to many industries: making materials lighter, stronger, and thinner; miniaturizing tools that can enter the human body to perform various tasks; and making compact electronics stronger and more powerful. Nanotechnology is often referred to as nanomanufacturing, because it entails manufacturing materials, systems, and products at the nanoscale.[6]

The National Nanotechnology Initiative (NNI) reports that nearly $100 billion will be invested in nanotechnology research in 2015, predominantly by the US government.[7] There has been a steady increase in funding since the 2000 level of $2 billion. Sales related to nanotechnology have been projected at nearly $50 billion in 2014 with nanomaterials accounting for nearly half of this total, followed by nanotools and a small portion from nanodevices.[8] The nanotechnology industry's production capabilities reach many industries including energy, biotechnology, electronics, and materials.[9]

After nanotechnology research breakthroughs, government and private funding, and growing revenue over the past decade, the nanotechnology industry has emerged from the Concept state and is currently nearing the state of Validation. In order for the industry to continue to emerge, the seven elements covered in this book need to continue to synchronize. Currently, firms are competing to establish standards, many technical applications are at or nearing viability, there is seed funding, supply networks are forming, production is being piloted, and there are early adopters. The elements are synchronized, so it is likely the nanotechnology industry will continue its emergence to the state of Validation. However, in order for the industry to reach the state of Diffusion, nanotechnology will need to continue to be deployed, markets and revenue will need to grow, firms and suppliers will need to lead, and regulations will need to protect intellectual property rights and encourage innovation.

6.4 Conclusion

The role and nature of production processes vary with the changing needs of firms in an emerging industry. Throughout the process of emergence, production is the expression or realization of an industry's ideas and inventions. The evolution of technology influences the nature of production, which in turn affects product value. For product value to increase, production processes and facilities must evolve to be capable of delivering the scale and reliability demanded by growing markets. This is in itself, however, dependent to some extent on investment in technologies and processes, which are in turn influenced by the market growth that is only possible through the evolution of the production base.

Production can influence the emergence of new industries when the scaling up of production enables the conversion of an idea or invention into a product. This evolution often begins with a prototype of an innovation, followed by a pilot version that can be used in order to collect performance and market data, and then, if emergence continues, the scaling up of production to batch or mass production. New methods of production, when synchronized with the other elements covered in this book, can also make new products possible and can spur the birth of new industries.

Exercises

- Describe how an idea or invention has progressed to become a product as it has moved from a prototype, to pilot testing, and then to scaled production.
- Identify a production process that has developed into an emerging industry and describe the elements that have influenced this progression.
- Choose an emerging industry and assess the interaction between production and other elements affecting the industry.

Notes

1 Drucker, P. (1985). *Innovation and entrepreneurship: Practice and principles*. New York, NY: Harper & Row.
2 Babbage, C. (1832). *On the Economy of Machinery and Manufactures*. London, UK: C. Knight, p. 7.
3 Marsh, P. (2012). *The New Industrial Revolution: Consumers, Globalization, and the End of Mass Production*. New Haven, CT and London, UK: Yale University Press.
4 *New York Times* (2012). A first: Organs tailor-made with body's own cells, September 15.

5 Available online at www.3ders.org/articles/20131014-gartner-3d-printing-to-result-in-100-billion-ip-losses-per-year.html. Gartner: 3D printing to result in $100 billion IP losses per year. March 15, 2015.
6 Nano Technology Initiative (2015). Available online at www.Nano.gov.
7 Ibid.
8 *BBC Research* (2015).
9 McDermott Will & Emery (2013). *2013 Nanotechnology Patent Literature Review*.

Further Reading

Babbage, C. (1832). *On the Economy of Machinery and Manufactures*. London, UK: Charles Knight.

Marsh, P. (2012). *The New Industrial Revolution: Consumers, Globalization, and the End of Mass Production*. New Haven, CT and London, UK: Yale University Press.

Pisano, G., & Shih, W. (2012). Does America really need manufacturing? *Harvard Business Review*, *90*(3), 94–102.

Weber, A. (1929). *Theory of the Location of Industries*. Chicago, IL: The University of Chicago Press.

7 Markets

Learning Objectives:

- Describe the role of markets in the birth and growth of industries
- Explain how visionaries affect the birth of new industries and early adopters drive industry emergence toward followers
- Understand the nature of demand for the growth of new industries
- Apply the concept of market gaps to explain the emergence of industries
- Analyze how various elements affect markets and how their synchronization affects industry emergence

Key Concepts:

- Visionaries
- Early adopters
- Followers

7.1 Introduction

Visionaries, early adopters, and eventually followers build markets for new industries. Growing demand helps drive firm strategy, technology development, investment, supply networks, production, and government action. When these elements move forward together, industries emerge and grow. For example, inventors, policymakers, and environmentalists envisioned a solar powered world, which helped spur the emergence of the solar PV industry due to new firms, advancing technology, increasing investment, expanding supply networks, production improvements, and government grants, subsidies, and procurement.

Market demand and corresponding supply help drive new industry growth, but emerging industries grow at an uneven pace and are frequently stalled by **market gaps**, which are market-related delays in the growth and emergence of an industry. New industries often begin with a technological breakthrough that is advanced by visionaries, or user–innovators, either via startup firms or as new divisions of firms in related industries.[1,2] Industries usually grow slowly as demand is limited to a small group of individuals (Visionaries, see Figures 1.1 and 7.1) who see possibilities before the majority of people. For example, the computer hobbyist community from the 1970s glimpsed the possibilities for the personal computer industry. A market niche can change as more people become aware of the technology (early adopters), and then economies of scale and technological developments lower prices and followers drive market demand. This process moves a product or service toward becoming the standard for a functional need: for example, automobiles became the dominant mode of transportation, and cell phones became the dominant communication device.[3,4,5] However, industry emergence can also stall due to market gaps in the progression. This chapter focuses on both the market drivers and deterrents of industry emergence and offers an understanding of how markets affect industry emergence and how demand forces affect the advancement of markets.

7.2 Markets and Emerging Industries

Markets influence industry emergence beginning with visionaries who see the possibility of a product or service, followed by early adopters who are keen to try products or services as soon as possible. Followers are next, once the product or service has been proven.

7.2.1 Visionaries

Visionaries are the first customers of a new industry, but they are much more because they are also user–innovators, or people with a vision for what a new industry could entail, and through their engagement they improve what the industry offers.[6] Visionaries are relevant for industry emergence because conceiving an idea or developing an invention is an essential beginning of a new industry. Growth from this idea or invention to a demonstrated product or service is often indirect and difficult. However, core to most visionaries is a strong belief in an alternative future. What visionaries and their corresponding new industries are up against can be described as a market interest gap (see Figure 7.1).[7,8] A **market interest gap** can arise from difficulties in generating demand for a new idea or invention. Such difficulties could stem from differences in tastes and preferences, moral concerns about the effects of new technologies, income levels of different market segments, and competing goods or services. For example, when electric vehicles were marketed

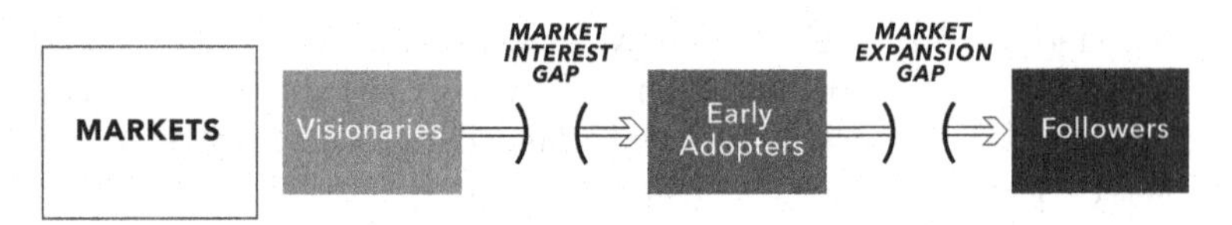

Figure 7.1 Markets and Industry Emergence.

in the 1990s, customer reception was lukewarm and remains so due to most people's comfort and preference for the driving range and fueling method of petroleum-based vehicles compared to inferior characteristics of electric vehicles. Visionaries in the electric vehicle industry are the hobbyists who retrofitted gasoline cars with electric motors and built early prototype electric vehicles.

Market interest gaps limit the availability of the resources necessary for a new industry to progress, including financing from any assortment of sources such as banks, government agencies, venture capitalists, and even friends and family.[9] The market interest gap may also be created by a mismatch between customer willingness to pay and supplier costs. For instance, when solar photovoltaic (PV) technology was introduced to the residential market its cost far exceeded customer willingness to pay. To help bridge this market interest gap, government entities can offer research funding and subsidies, and angel investors can be key external partners at this juncture, but other potential stakeholders, such as visionary customers, can help provide a bridge across the market interest gap in the form of pre-orders for to-be-completed technology, products, and services,[10] for example with crowd-sourced funding via Kickstarter or Indiegogo.

7.2.2 Early Adopters

Early adopters are customers who purchase a new product or service as soon as it is available for sale (see Figure 7.1).[11] They are different from visionaries in that they are not user–innovators, but instead they see the value in the product or service and are willing to pay for it via a market exchange.[12,13] They are part of an "exchange of commodities at various prices between two types of economic agents—consumers and firm."[14] Customers for a product or service seek value from a purchase that meets or exceeds what they have to pay for obtaining the product or service. Customers may make purchases because they receive the value from products or services in the form of food and health care, of enjoyment, or because they receive profits by combining the purchase as an input to a product or service they in turn can sell. Their willingness to pay is the highest price they would choose to pay for a particular good or service, considering the value they anticipate receiving from the purchase. Customers' willingness to pay is influenced by several factors, and in turn, these factors influence whether or not and when someone adopts a product or service.

Customers' tastes and preferences affect their willingness to pay for a product or service relative to competitors' offerings. Tastes and preferences are affected by culture, information, and exposure to a product or service via personal networks and advertising. For example, which social media venues a person chooses: Facebook, Snapchat, Instagram, Vine, Tumblr, among others, is influenced by a person's age, as well as their peer network.

Moral concerns may affect people's support for the products or services of an emerging industry. If people feel human rights could be infringed upon they may resist or even protest the emergence of an industry.[15] Only after these concerns are addressed is the market interest gap likely to be bridged. For example, there has been resistance to the recreational use of UAVs due to concerns about the invasion of privacy. Ongoing development of regulatory measures is aimed at addressing these concerns, which may help bridge this industry's market interest gap and enable further emergence of the industry.

Income level affects customer demand, as in most cases, customers' willingness to pay increases with wealth. In addition, once basic needs are met, higher income levels influence customers to expand the features they consider when making a purchase. Income levels play an important role when defining market segments, as this enables sellers to identify different groups of customers. Early adopters of products or services are often customers with higher income levels.

Competing goods and services define the scope of choices for customers by putting value in context. Substitutes offer functional options, such as electric vehicles offering a transportation option to gasoline or fuel cell vehicles. Having options tests customers' calculation of value and their willingness to pay. In addition, complementary goods or services can affect a customer's willingness to pay, when owning one product increases a customer's likelihood to purchase another. For example, owning solar panels may increase a customer's demand for an electric vehicle because one can complement the other. In whole, early adopters purchase products or services when enough of these factors are met satisfactorily. The following is an emerging industry case study on the electronic commerce industry that illustrates shifts in market demand patterns caused by an emerging industry.

Emerging Industry Case Study: Electronic Commerce

The **electronic commerce industry** entails buying and selling goods and services via the Internet. Physical goods are packaged and then shipped to buyers or content (e-books, music, etc.) is made available for download. Services such as legal and tax assistance or graphic design can be purchased online, or services can be paid for and scheduled online but delivered in person (e.g. cleaning services, tutoring).

The electronic commerce (e-commerce) industry has grown over the past 50 years from nearly no sales in the 1960s to over $1.4 trillion by 2015.[16] The roots of the e-commerce industry are the development of Electronic Data Interchange in the 1960s, which enabled companies to share business documents such as invoices, order forms, and shipping confirmations with other companies via computer-linked intranets. Also in the 1960s, the US military developed ARPANET to ensure that crucial communications were circulated in the event of a nuclear attack. The original ARPANET connected four large US research universities, and was used for informal commerce by students at these universities. In the early 1980s, individual computer users were able to send e-mails, participate in Listservs and newsgroups, share documents, and even buy and sell through networks like BITNET and USENET. Visionaries drove the e-commerce industry during the first 25-year period of industry emergence as they were the first customers, but they were also the first user–innovators who experimented and improved electronic commerce.

CompuServe offered an online sales platform called the Electronic Mall in 1984, which enabled hundreds of merchants to sell directly to consumers. In 1995, Amazon.com started selling books online, and eBay launched an online auction site. These companies offered early adopters of e-commerce the opportunity to make online purchases. The past 20 years have witnessed phenomenal growth in e-commerce, as Amazon now sells most everything, and along with companies such as JD.com, eBay, and Alibaba has multi-billion dollar annual revenue. Over 80 percent of Internet users have made at least one online purchase,[17] indicating that the e-commerce industry has transitioned to followers and is an established venue for buying and selling.

In addition to phenomenal growth, the e-commerce industry's emergence has changed consumer shopping patterns in the traditional retail industry. Spending habits such as holiday shopping have changed as sales have smoothed out throughout the year, reducing typical holiday spending by 50 percent in traditional brick and mortar retail stores from 2010 to 2013.[18] Shoppers have also adjusted their browsing patterns. They browse physical stores less, using online insight to help them figure out what they want and then focusing their trips to physical retailers to purchase items from retailers offering the best price. Shoppers visited an average of five stores per mall trip in 2007, and by 2014 they only visited three stores per trip.[19] The e-commerce industry has caused shifts in willingness to pay due to an increase in options, and there are likely to be lower promotion and distribution costs due to the reduced need for physical stores.

7.2.3 Market Followers

Market followers decide to purchase only after the product or service has proven to be more fashionable, higher quality, or less expensive than was originally perceived. Followers are hindered by a **market expansion gap**, which often exists between demonstration of a product or service and growth opportunities and is represented by the challenge of moving from niche markets to growth markets, which allows for economies of scale and competitive pricing (see Figure 7.1). Just as with the market interest gap, the market expansion gap may need external actors to provide bridging and often can entail large infusions of capital and a more developed and supportive supply network. Larger amounts of money from a financial partner are often necessary for moving from a demonstration or pilot facility to full production, but contract manufacturing partnerships, licensing agreements, and government procurement can help move an industry forward across the market expansion gap.[20,21] An example of an emerging industry hindered by a market expansion gap is the biofuels industry. The challenges of scaling production (lack of local and affordable feedstock, limited distribution, etc.) have limited the industry to primarily local, niche markets. Building manufacturing capability and a supply chain are two initiatives firms have tried in order to help the emergence of the biofuels industry.

If an industry continues to emerge, the focus of the industry shifts to pricing, booking orders, obtaining shelf space, and establishing a distribution network. Price competition and limits to distribution are common challenges for an industry as it continues to emerge. Moving beyond a locally successful introduction or temporary subsidy-driven entry can be critical for strengthening and sustaining a market position. Market growth opportunities may require several well-placed alliances in the case of locational expansion and long-term partnerships for temporal expansion. The market gaps—interest and expansion—covered above affect the emergence of industries. Bridging these market gaps, often via alliances, is essential for new industries to form, grow, and be sustained. The following is an emerging industry case study on the solar photovoltaic industry that illustrates the challenges market gaps pose for an emerging industry.

Emerging Industry Case Study: Solar Photovoltaics

The solar photovoltaic (PV) industry offers an illustrative example of industry emergence as shown in Figure 7.1. The market gaps and bridges referred to conceptually above are demonstrated for the solar PV industry through time periods in the emergence of the industry.

Solar PV technology progressed slowly due to technology limitations and a market interest gap. Antoine-Cesar Becquerel, the original visionary, first discovered the direct relationship between sunlight and electricity, or the photovoltaic effect, in 1839.[22] Photovoltaic devices generate electricity directly from sunlight via an electronic process that occurs naturally in some materials. Electrons in certain types of crystals are freed by solar energy and can be induced to travel through an electrical circuit, powering electronic devices. The first development of diodes occurred in 1938, and transistors were developed in 1948, enabling the building of a solar cell.[23] Bell Labs patented the first solar cell based on silicon in 1954, and Hoffman Electronics led the development of more efficient solar cells. Sharp Corporation began researching photovoltaic cells in the 1950s and succeeded in producing practical silicon PV modules by 1963.[24]

Technological breakthroughs and entrepreneurial ventures highlight the beginning of the emergence of the solar PV industry. Scientists (starting with Becquerel), government-funded labs (Bell Laboratories), startups (Hoffman Electronics), and entrepreneurial divisions of large corporations (Sharp Corporation) were all significant visionaries for the initial emergence of the solar PV industry. However, limited demand caused a market interest gap and slowed the emergence of the industry (see

Figure 7.1). This gap did not begin to close until interest grew from the US space program and the environmental movement in the 1960s. The alliances between scientists, government labs, firms, and the US space program increased technology development, reduced risk and financial exposure, and improved focus on solar PV technology. All of these elements had to come together, or be synchronized, in order for solar PV technology to be advanced to the point of providing a viable power source. These alliances were necessary to begin to bridge the sizeable market interest gap between the understanding of PV technology and its refinement for viability and early adoption. The solar PV case shows the dual importance of technology development-based alliances (to advance the idea/invention) and demand-based alliances (to focus the technology application) during the birth and early growth of an industry.

Once solar PV technology was developed to the point of demonstration, interest grew because of the unlimited nature of the fuel source, the environmental movement, and the efforts of organizations and their alliances profiled above. However, solar PV power remained only a small portion of installed renewable energy capacity. A market expansion gap existed as the solar PV industry stalled, and cost hurdles slowed the adoption of the technology. These cost hurdles were exasperated by the continued availability of less expensive alternatives for electricity generation, including renewable energy sources such as wind power, and natural gas as the most competitive alternative.[25]

New business models are helping bridge the market expansion gap (see Figure 7.1). High up-front capital costs are a primary deterrent to the adoption of solar PV technology. Solar PV technology providers have installed PV systems on customers' roofs and negotiated long-term electricity purchase agreements. Solar PV providers targeted universities, hospitals, governments, and similar stable institutions with this business model.[26] This allowed the solar PV providers to overcome the up-front capital challenge, develop a steady, long-term revenue stream, and increase the adoption of solar PV technology. Providers have also developed cost-share models with municipal governments where both share a portion of the up-front capital costs for residential solar installation, and this cost is included in property taxes. These new business models have required new types of alliances and have helped increase the scale of production of solar PV, lowering per-unit production costs and

(*Continued*)

helping to bridge the market expansion gap, aiding the continued emergence of the solar PV industry. The solar PV industry case shows the importance of alliances in order to advance an emerging industry from early adopters to followers, as these types of alliances have helped reduce production and installation costs and risks.

The market expansion gap has begun to close, as demonstrated by the over 30 percent annual growth in installed solar PV capacity since 2000 (see Figure 7.2).[27] This market gap has been closing due to alliances between multiple entities connected to the solar PV industry. Firms have continued to enhance their supply chains in order to reduce costs, and governments have offered subsidies in the form of research funding, guaranteed loans, price supports, and procurement.[28]

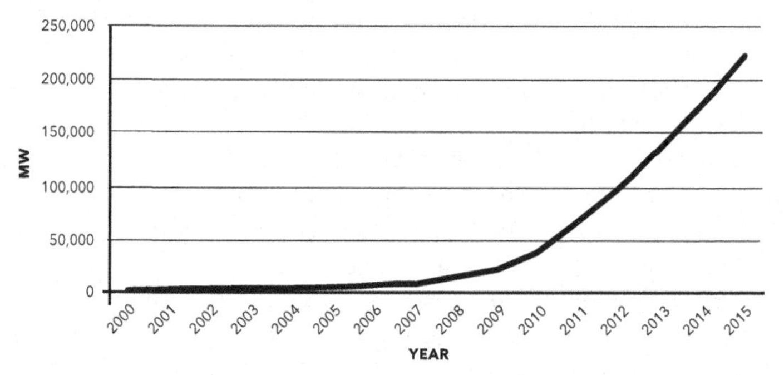

Figure 7.2 Global Solar PV Electricity Generation.[29]

Government intervention in Japan, Germany, Spain, the US, and the UK has been critical in helping solar PV electricity to be more cost competitive. Startup firms and multinational corporations from related and supporting industries have helped drive the industry across the market expansion gap, and the market growth opportunities have been partially realized by the evolution of component and complementary industries such as silicon, electric utilities, and construction. Bridging each of the gaps in industry emergence (market interest and market expansion) has depended on alliances, though different types of alliances have played larger roles at different times, and many types of alliances are likely to be necessary for the solar PV industry to continue to emerge. While rising demand for energy and concerns about environmental protection and energy security motivate continuing demand for solar

PV technology, alliances between government, investors, firms, and consumers, and synchronization of the elements highlighted in this book will be needed for the continued emergence of the solar PV industry.

7.4 Conclusion

Visionaries, early adopters, and eventually followers are the customers responsible for helping new industries emerge. Their growing demand helps drive firm strategy, technology development, investment, supply networks, production, and government. When these elements move forward together, industries emerge and grow. However, market gaps can delay the emergence of industries in several ways. This chapter has explained and illustrated market gaps (interest and expansion) and shown how market gaps can slow and even stall the pace of industry emergence. This chapter also emphasizes the importance of multiple elements and their synchronization for industries to emerge. Technological breakthrough is critical for the birth and growth of new industries, and is in turn dependent upon market demand. Likewise, as shown in the solar PV example, market demand is often dependent upon elements such as firm strategy, investment, the support of supply networks, production advancements, and government mechanisms. Understanding how these elements interact is the theme of this book and is essential for understanding emerging industries.

Exercises

- Choose an industry and explain the timing and roles of visionaries, early adopters, and followers in the emergence of this industry.
- Describe the circumstances that may cause market interest and market expansion gaps in emerging industries.
- Which elements affecting the solar PV industry need to be synchronized in order for the industry to continue to emerge?

Notes

1 Malerba, F. (2007). Innovation and the dynamics and evolution of industries: Progress and challenges. *International Journal of Industrial Organization*, 25, 675–699.
2 von Hippel, E. (2005). *Democratizing Innovation*. Cambridge, MA: MIT Press.
3 Rogers, E. (2003). *The Diffusion of Innovation*. New York, NY: Simon & Schuster.

4 Cohen, S. (2010). Innovation-driven industry life cycles. In V. K. Narayanan, & G. Colarelli O'Connor (Eds.), *Encyclopedia of Technology and Innovation Management*. New York, NY: John Wiley & Sons Ltd.
5 Funk, J. (2012). *Technology Change and the Rise of New Industries*. Palo Alto, CA: Stanford University Press.
6 von Hippel, E. (2005). *Democratizing Innovation*. Cambridge, MA: MIT Press.
7 Jolly, V. (1997). *Commercializing New Technologies: Getting from Mind to Market*. Boston, MA: Harvard Business School Press.
8 Dodgson, M. (2000). *The Management of Technological Innovation: An International and Strategic Approach*. Oxford, UK: Oxford University Press.
9 Moore, G. (2002). *Crossing the Chasm: Marketing and Selling High-tech Products to Mainstream Customers*. New York, NY: Harper Publishing.
10 von Hippel, E. (2005). *Democratizing Innovation*. Cambridge, MA: MIT Press.
11 Rogers, E. (2003). *The Diffusion of Innovation*. New York, NY: Simon & Schuster.
12 Linton, J., & Walsh, S. (2004). "ntegrating innovation and learning curve theory: An enabler for moving nanotechnologies and other emerging process technologies into production. *R&D Management*, *34*(5), 517–526.
13 Suarez, F. (2004). Battles for technological dominance: An integrative framework. *Research Policy*, 33, 271–286.
14 Spulber, D. (1989). *Regulation and Markets*. Cambridge, MA: MIT Press.
15 Vallor, S. (2016). *Technology and the Virtues: A Philosophical Guide to a Future Worth Wanting*. Oxford, UK: Oxford University Press.
16 Available online at www.cpcstrategy.com/blog/2013/08/ecommerce-infographic/ (accessed December 20, 2015).
17 Available online at www.cpcstrategy.com/blog/2013/08/ecommerce-infographic/ (accessed December 20, 2015).
18 Banjo, Shelly, and Drew Fitzgerald (2014). Stores confront new world of reduced shopper traffic. *Wall Street Journal*, 16 Jan. 2014. Web: 09 May 2015.
19 Ibid.
20 Stuart, T. E. (1988). Network positions and propensities to collaborate: An investigation of strategic alliance formation in a high-technology industry. *Administrative Science Quarterly*, 668–698.
21 Todeva, E., & Knoke, D. (2004). Strategic alliances and models of collaboration. *Management Decision*, *43*(1), 123–148.
22 Bradford, T. (2007). *The Solar Revolution: The Economic Transformation of the Global Energy Industry*. Cambridge, MA: MIT Press.
23 Ibid.
24 Ibid.
25 Lorenz, P., Pinner, D., & Seitz, T. (2008). *The Economics of Solar Power*. Chicago, IL: McKinsey & Co.
26 Haley, U., & Schuler, D. (2011). Government policy and firm strategy in the solar photovoltaic industry. *California Management Review*, *54*(1), 17–38.
27 IEA (2013). *Trends in Photovoltaic Applications*.
28 Harborne, P., & Hendry, C. (2012). Commercialising new energy technologies: Failure of the Japanese machine? *Technology Analysis & Strategic Management*, *24*(5), 497–510.
29 Adapted from IEA (2013). *Trends in Photovoltaic Applications*. Available online at http://iea-pvps.org/fileadmin/dam/public/report/statistics/FINAL_TRENDS_v1.02.pdf.

Further Reading

Dodgson, M. (2000). *The Management of Technological Innovation: An International and Strategic Approach.* Oxford, UK: Oxford University Press.

Jolly, V. (1997). *Commercializing New Technologies: Getting from Mind to Market.* Boston, MA: Harvard Business School Press.

Moore, G. (2002). *Crossing the Chasm: Marketing and Selling High-tech Products to Mainstream Customers.* New York, NY: Harper Publishing.

Rogers, E. (2003). *The Diffusion of Innovation.* New York, NY: Simon & Schuster.

Vallor, S. (2016). *Technology and the Virtues: A Philosophical Guide to a Future Worth Wanting.* Oxford, UK: Oxford University Press.

von Hippel, E. (2005). *Democratizing Innovation.* Cambridge, MA: MIT Press.

8 Government

Learning Objectives:

- Gain an understanding of the roles of government and the variety of government mechanisms affecting emerging industries
- Learn how to analyze the effects of types of government mechanisms on emerging industries
- Assess the effects of other elements such as firm strategy, investment, and markets on government mechanisms

Key Concepts:

- Infrastructure
- Research funding
- Procurement
- Regulation

8.1 Introduction

Government plays a wide range of roles in society and affects business in many ways, including influencing the emergence of new industries. The primary roles of government can be categorized as infrastructure provider, funder, customer, and regulator. Government can play these roles using a number of mechanisms to influence the emergence of new industries, and government can use these mechanisms at different times for different effects.[1] Governments maintain infrastructure that helps build an environment for industries to begin to emerge. Governments can use grants, subsidies, and procurement as an industry continues to emerge, and can use regulations as an industry gains momentum and becomes established.[2] This chapter describes the roles of government and the variety of government mechanisms affecting emerging industries, and assesses the effects of other elements such as firm strategy, technology, investment, and markets on government mechanisms.

8.2 Government and Industry Emergence

Government often helps establish the foundation for the birth and growth of new industries. Government can influence the direction and pace of industry growth, and a variety of mechanisms are used at different times for different effects. As an industry first emerges, government's role is most likely to be the development of basic research through the support of infrastructure and technology development. After ideas and technology start gaining traction, government uses grants, subsidies, and procurement contracts to help encourage industry emergence. Government also regulates industries with mechanisms such as laws, standards for goods and services to be sold, and food and drug approval.

8.2.1 Infrastructure Provider

Government provides infrastructure, which sets the foundation for the birth and growth of new industries (see Figure 8.1). **Infrastructure**, which is the basic structures and facilities needed for the operation of society, can include roads, railways, and airports, along with communications and utilities such as energy and water. Infrastructure can also include education, training, health, welfare, and the legal environment. In almost all countries, the provision of infrastructure is seen as the duty of government. Individual tasks such as construction and operation may be devolved to private companies, but the responsibilities remain with government.

Perhaps the most vital form of infrastructure for encouraging emerging industries is education. The educational level of a country's citizens and its potential for more complex and value-adding economic activities is likely to aid technology innovation and industry emergence. The educational infrastructure involves not only the education process itself but also the environment of examinations and institutions designed to motivate, reward, and monitor progress and capability individually and collectively.

The legal environment is also a critical aspect of the infrastructure for aiding industry emergence. An independent judiciary, separate from government but enabled by it, allows research, development, and commercialization activities to proceed in an orderly manner. From the protection of intellectual property, to the binding nature of contracts, to

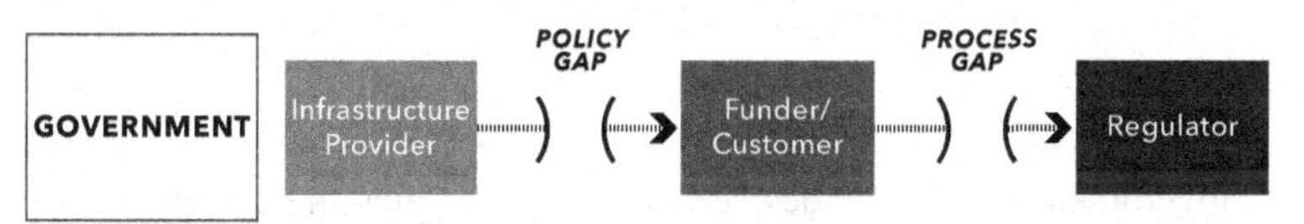

Figure 8.1 Government and Industry Emergence.

protection against fraud and wrongdoing, the legal infrastructure allows technology development and industry emergence to proceed in such a way that disputes will be handled in a fair and orderly manner and the distorting effects of corruption and favoritism can largely be avoided, or at least controlled. **Intellectual property** is something unique that is created. Ideas alone are not intellectual property—they must be realized. Government establishes the rules for protecting intellectual property and supports a legal system as a venue for defending property rights. The purpose of intellectual property protection is to encourage progress. This is especially relevant for emerging industries, as technological innovation is encouraged when inventors believe their work will be protected for their commercialization.

However, the infrastructure needed for new industries is not always enough for industries to continue to grow. Sometimes there will be a **policy gap** preventing government support from extending from the provision of infrastructure in general to funding and procurement for a particular emerging industry (see Figure 8.1). Government funding and procurement policy needs to be industry-specific for it to be most effective in encouraging an industry to continue its emergence. This policy gap is also present if government does not maintain its support consistently. Therefore, governments can bridge the policy gap by setting clear policy for funding and procurement support that is specific to the desired emerging industry and by maintaining long-term support of the industry as it emerges to the state of diffusion. Government needs to present a vision of which emerging industries are most promising for the country's economic development. This vision is usually captured in industrial policy statements or documents. Government agencies commission studies using tools such as scenario planning, road-mapping, and Delphi processes in order to assess the current state of industry emergence, the gaps preventing industries from growing, the resources needed to bridge the gaps, and the potential impacts from government mechanisms. Example efforts include the UK's Foresight program, critical technology reports in the United States, and the World Economic Forum's regional and country scenarios. Government will often promote its industrial policy through briefings, reports, conferences, and technical workshops. If a government bridges the policy gap and provides a long-term vision for support of an emerging industry, it is more likely to continue its support as a funder and customer.

8.2.2 Funder and Customer

Government often provides seed funding that aids the growth of new industries. Examples include the UK Research Council funding and the US Small Business Innovation Research (SBIR) program. Many technologies need long-term, sizeable, and patient sources of funding in order for research, discovery, and viability. These funds are usually competitively awarded to universities, laboratories, and startup firms. Government also funds specific technologies for supporting new industries, for example electric vehicle charging stations for electric vehicles, communications infrastructure for electronic commerce and social networking, and national laboratories for robotics, virtual reality, and biotechnology.

Government may also be a primary customer for the products of an emerging industry, especially as early adopters. Procurement contracts, often of sizeable value, can offer a new industry the demand needed to improve technology and expand supply networks and production to approach economies of scale.

Once technology has been developed and it is focused on commercialization, standards and regulatory approval become more important. This is where a process gap can appear and stall industry emergence. A **process gap** is the delay between research and development and commercialization, and can come in the form of the slow approval of a drug or device, or in the form of political opposition to the promotion of an industry, such as the emerging UAV industry. If the process gap is bridged either through persistence or shifting political support, government continues with its roles as infrastructure provider, funder, and customer as an industry grows, but the role of government also expands to the establishment of regulations and the setting of standards (see Figure 8.1).

8.2.3 Regulator

Government's primary role for many industries is as the regulator, and this is usually the case for an emerging industry, especially after government-provided infrastructure, funding, and procurement have helped with the birth and initial emergence of a new industry. Government introduces regulations for a variety of reasons, from ensuring safety to controlling unfair practices. Safety is widely seen as one of the main duties of government, especially for new, often experimental technology, which is the basis for most emerging industries. Approval of products can be mandatory before they can be sold in the market, for example, with the US Food and Drug Administration (FDA) for the pharmaceutical industry. The following emerging industry case study of the virtual reality industry demonstrates how the roles and mechanisms of government can influence how an industry emerges.

Emerging Industry Case Study: Virtual Reality

Virtual reality simulates experiences in real places or imagined places and enables the user to interact in these places. Virtual reality creates artificial experiences for the senses, including sight, hearing, and touch. Simulations are shown on computer screens or stereoscopic displays with some systems including tactile experiences with wired gloves or treadmills. The primary uses are for medical, military, and gaming applications, such as surgical, pilot, or combat training and virtual presence gaming.

Government, acting as an infrastructure provider, funder, customer, and regulator, has affected the virtual reality industry in many ways as this industry has emerged. Government has developed educational infrastructure through student support in subjects such as computer science, industrial arts, and engineering. Government has also protected intellectual property, funded research and startup companies, and assessed and regulated the safety of new technologies.

Government has provided research funding throughout the emergence of the virtual reality industry. Beginning in the 1960s, the US government invested in computer modeling, visualization techniques, and flight simulators for military, civilian, and university research as precursors to virtual reality technology.[3] These investments and the technology they helped fund have laid the foundation for the emerging virtual reality industry. The virtual reality industry has continued to emerge and private investment has increased, while government funding has continued for complementary technology such as head-mounted displays and advanced simulators. Federal funding also supports university research centers and talent development, as graduate students and academic researchers who receive government support contribute to the advancement of technology and support leading companies in the industry.[4]

Government further supports the virtual reality industry infrastructure by protecting and incentivizing developers of intellectual property. For example, as the virtual reality industry began its emergence around 1990, the number of virtual reality-related patents filed in the US per year began to climb, peak in 2008, and then decline (see Figure 8.2).[5] The decline is likely due to firm consolidation in the industry as technology standards were approached and a group of leading firms made acquisitions, while weaker firms disappeared.

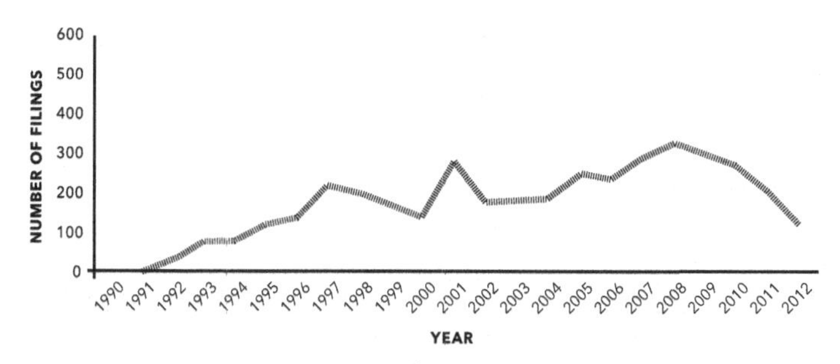

Figure 8.2 Virtual Reality Patents.[6]

Government aids the emergence of the virtual reality industry through procurement, namely for space (NASA), military, and medical simulation training. NASA projects include virtual environment simulation for space travel and exploration of environments such as Mars and other planets. Military training enables warzone simulation for building tactics, techniques, and teamwork under different types of combat situations, scenarios, and terrains.[7] Air Force pilots have also been early adopters and continue to benefit from and advance the virtual reality industry by training with flight simulators. Medical simulation is used to train surgeons, emergency medical technicians, medical students, and nurses. This enables medical personnel to be trained to deal with a wider variety of medical situations, and trainees can practice repeatedly in order to refine their skills.

Government has played a significant role in the emergence of the virtual reality industry and has interacted directly with technology, provided and supported markets, and contributed substantial investment. In addition, government has helped guide the industry by funding standard-setting research and companies and by protecting specific technology regimes. Support for firms through grant and loan programs, for example SBIR, and support for firms producing virtual reality equipment and related and supporting products and services, for example stereoscopes and software, has also helped the virtual reality industry emerge.

8.3 Government and Synchronization

Government usually helps establish the foundation for the birth and growth of new industries. This foundation may be the educational infrastructure, funding for research, as a lead customer, regulations, and technical and financial support for production capabilities and supply networks.

While government entities can play a significant role in emerging industries, they and other entities can introduce uncertainty as policies can change. While many stakeholders will often seek to influence government, the uncertainty of emerging industries requires firms and investors in particular to enhance their ability to recognize the variability in an industry and their capability to adjust their strategy when industry change dictates adjustment. For example, the role of government is particularly important in the case of electric vehicles. Governments in the US, Israel, Denmark, Norway, Japan, China, Germany, and the UK have introduced incentives for electric vehicle purchases, and billions of dollars have been dedicated to promote electric vehicle manufacturing and advancements in support technologies such as batteries, power electronics, and charging stations. These subsidies have included help to locate electric vehicle manufacturing, loans for electric vehicle manufacturers, and grants for battery and component firms and battery recycling. Governments have also procured electric vehicles for test fleets, built charging systems, and upgraded power grids. Consistent and long-term government support is necessary for industries such as electric vehicles to emerge and grow to sustainability.

Multiple elements come together for an industry to emerge, and government is usually in the midst of these interacting elements. For example, the emergence of the biofuels industry is the result of a combination of firm strategy, technological innovation, markets, and government intervention. Government mechanisms such as subsidies, procurement, investment incentives, public education programs, regulations and standards, and macroeconomic policies have encouraged innovation and commercialization, leading to biofuels industry growth and tax revenues. The following emerging industry case study of the biotechnology industry also demonstrates how government can influence how an industry emerges.

Emerging Industry Case Study: Biotechnology

Biotechnology uses biological processes in the development or production of a product or in the technological solution to a problem. The biotechnology industry is a science-based industry as it "attempts to both create science and to capture value from it,"[8] and government mechanisms play major roles in the biotechnology industry's emergence. The first biotechnology firm was Genentech, which was founded in California in 1976. Figure 8.3 shows the emergence of the biotechnology industry based on its global sales growth from its beginning to over $130 billion in 2013.[9] However, while revenue has grown steadily, profits have been limited due to immense research and development expense and risk, with only a few firms showing positive earnings.

There are over 5,000 biotechnology firms in the world, with significant clusters of activity in San Francisco, Boston, and San Diego in the US, and in Cambridge, UK, Basel, Switzerland, and Singapore. These clusters have developed in part because of the targeted efforts of national and regional governments choosing biotechnology as a strategic industry for jobs and economic development. Governments support the locational development of the biotechnology industry through infrastructure funding for education and training, investment in research laboratories and for startup firms, and development of standards and regulations to ensure safety and quality and protect property rights. Most of the firms in the biotechnology industry are small, and the barriers to enter the industry are high due to the need for groundbreaking research and sizeable amounts of money to fund drug development. Early funding is often in the form of government research grants. Biotechnology firms are also highly specialized, which increases their vulnerability should a rival beat them to government approval or to market with a drug.

Biotechnology firms are primarily valued based on their possession of intellectual property. Supply networks are highly specialized, providing research and testing equipment, materials, and highly trained people. Many small biotechnology firms have little or no access to distribution, which often necessitates the licensing or sale of their approved drugs to larger and better-funded

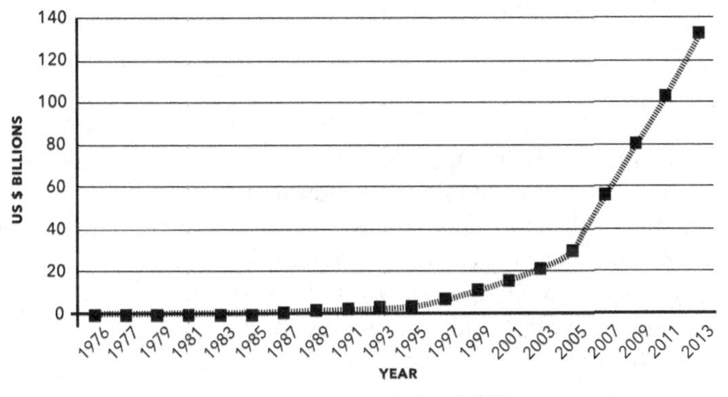

Figure 8.3 Global Biotechnology Sales Growth.[10]

(*Continued*)

pharmaceutical companies. Governments and large hospital groups often hold significant bargaining power with biotechnology firms because their products are expensive and highly specialized, and in the case of single-payer systems, there can be many people who use the drug, but few paying customers.

A competitive challenge for biotechnology firms is the production of generic drugs. Firms spend millions of dollars to discover and create a new drug, which they must sell at a high price in order to recover the research and development costs. In countries with limited government control, generic drug makers copy the drug formulas and sell them for a fraction of the cost incurred by the original biotechnology firm. Therefore, enforcement of appropriation rights is an important role for government to support the biotechnology industry.

8.4 Conclusion

Government affects emerging industries through the establishment of a foundational infrastructure, which can encourage the birth of new industries. Government also affects industry emergence through direct investment in research and technology, and through procurement, subsidies, and regulation. Government can influence supply through funding, infrastructure, and skills, and demand through procurement. Moreover, government sets rules as the regulator of competition, standards, and safety. Government action is unique among the elements of emerging industries in that the intention is usually to influence the industry, perhaps as a whole. Other elements essentially involve agents with their own interests uppermost in their minds. Government action is broad reaching, influences the other elements covered in this book, and is often the catalyst for industry emergence.

Exercises

- Choose an industry and describe how different government mechanisms have played a role as the industry has emerged.
- Compare the motivation of government vs. the motivation of other entities influencing emerging industries.
- Describe how government is affected by other elements such as firms, technology, investment, supply networks, production, and markets.

Notes

1 Hung, S., & Y. Chu. (2006). Stimulating new industries from emerging technologies: Challenges for the public sector. *Technovation*, *26*(1), 104–110.
2 Harborne, P. and Hendry, C. (2012). Commercialising new energy technologies: Failure of the Japanese machine? *Technology Analysis & Strategic Management*, *24*, 497–510.
3 National Research Council (1999). *Funding a Revolution: Government Support for Computing Research*. Washington, DC: The National Academies Press.
4 Ibid.
5 Greenbaum, E. (2014). A virtual reality patent landscape analysis, blog post.
6 Adapted from Greenbaum, E. (2014). A virtual reality patent landscape analysis, blog post, Available online at https://greenbaumpatent.wordpress.com/2014/01/23/a-virtual-reality-patent-landscape-analysis/ (accessed January 23, 2014).
7 Dourado, Antônio O., & Martin, C. A. (2013). New concept of dynamic flight simulator, Part I. *Aerospace Science and Technology*, *30*(1), 79–82.
8 Pisano, G. (2006). *Science Business: The Promise, the Reality, and the Future of Biotech*. Boston, MA: Harvard Business School Press.
9 Available online at www.bbbiotech.ch/en/bb-biotech/ (accessed February 2, 2016); Available online at www.ibisworld.com/industry/global/global-biotechnology.html (accessed February 2, 2016).
10 Adapted from and available online at www.bbbiotech.ch/en/bb-biotech; www.ibisworld.com/industry/global/global-biotechnology.html.

Further Reading

Block, F., & Keller, M. (Eds.) (2010). *State of Innovation: The U.S. Government's Role in Technology Development*. Boulder, CO: Paradigm Publishers.

Harborne, P. and Hendry, C. (2012). Commercialising new energy technologies: Failure of the Japanese machine? *Technology Analysis & Strategic Management*, *24*, 497–510.

Hung, S., & Chu, Y. (2006). Stimulating new industries from emerging technologies: Challenges for the public sector. *Technovation*, *26*(1), 104–110.

Livesey, F. (2013). How should governments support the emergence of new industries in leading economies? *International Journal of Public Policy*, *9*(1–2), 108–130.

9 Synchronization for Industry Emergence

Learning Objectives:

- Explain the interaction of elements affecting emerging industries
- Understand how synchronization affects industry emergence

Key Concepts:

- Element interaction
- Synchronization

9.1 Introduction

Synchronization, or the coordination of multiple elements so that they reach a particular state all at the same time, is the key concept that enlightens us on industry emergence. Industries are born and grow when key elements are coordinated in time, or are in sync. This chapter focuses on the interaction of the elements influencing industry emergence described in the preceding chapters and elaborates on the concept of synchronization and its importance for the emergence of new industries. Element interaction and synchronization are related concepts because the interaction of the elements can result in co-development and coordination leading to synchronization. The seven elements—firm strategy, technology, investment, supply networks, production, markets, and government—affect each other as they interact and develop, in some cases driving each other forward and in other cases restricting each other. In this chapter, the interaction between the elements is explored through analysis of the birth and growth of new industries. We illustrate these interactions with example emerging industries and data representing the seven elements, and we elaborate on the synchronization of the elements, which influences new industries' movement from the Concept, to the Validation, and to the Diffusion states of industry emergence.

9.2 Element Interaction

For a new industry to emerge, a technology, or usually a number of technologies and resources, (sometimes referred to as complementary assets) must be incorporated into products.[1,2] These products are offered to the market in the hope of sales, thus generating returns for the businesses and investors. The response of the market depends on the attractiveness of the benefits offered by a product, given the price at which it is sold.[3] Initially, these products will be attractive to a relatively small number of potential customers, for example visionaries, then early adopters, and then followers,[4] for whom the benefits to be derived from the product outweigh the cost associated with it (cost in financial terms, as well as in terms of factors such as risk or unreliability). Initial sales, however, can provide returns that become a source of investment to further develop the product. Furthermore, evidence of sales might encourage others to provide investment funds.

The emergence of a new industry relies on a positive feedback cycle to enable the development of the key elements, all of which are interdependent. Increased market demand may lead to increased availability of funding, which in turn enables the development of technology, which can be integrated to increase the benefits or reduce the cost of the products offered to the market, which could lead to increased demand, and so on. Technology development influences the nature of production, which in turn affects product value. For product value to increase, production processes and facilities as well as supply networks must evolve to be capable of delivering the scale and reliability demanded by the growing markets.[5] This is in itself, however, dependent to some extent on investment in technologies and processes, which is in turn influenced by market growth, which is only possible through the development of the production base and supply networks. Government's role changes as an industry emerges, first aiding emergence with infrastructure for discovery, then helping with funding and procurement, followed by regulations and standard-setting. The interaction of the elements act as a feedback cycle, however, this interaction can also operate to constrain or even reverse the growth of an industry. Falling demand can result in reduced availability of investment funds, and hence constrain the development of technology, leading to a lack of competitiveness in terms of product value and further reduced sales.

These interactions usually start with the discovery of an idea or technology. Industry growth can be hindered by what are described as gaps, for example knowledge gaps, market interest gaps, etc. A knowledge gap is caused by missing knowledge or complementary technology, and a market interest gap is a situation that occurs when limited market appeal restricts the advancement of technology.[6,7] These gaps can be formidable, and some technologies never advance beyond the point of discovery because they are too expensive to develop, lack commercial application, or are superseded by a competing technology.[8] This was the case with solar photovoltaic technology, which languished in laboratories until US government-driven need for remote power for the space program spurred further development of the technology. In this case, the incentive to invest in the development of the core technology arose from a clear market demand (the space program), and the availability of government investment funding that accompanied it. The resulting improved product value subsequently enabled increased market demand, at least partly by overcoming the resistance to adoption, generating increased investor demand, and creating the possibility of further development of the technology.[9]

The following sections offer more detail on the growth of firms, technology, investment, supply networks, production, markets, and government during the emergence of several, illustrative industries.[10] After the sections on each of the seven elements, the elements are joined in one figure (Figure 9.8) to highlight element interaction and synchronization as industries emerge and grow.

9.2.1 Firm Strategy

Assessment of the industry structure of an emerging industry revolves around the strategic behavior of firms and the interactions between the firms and other elements such as customers, investors, and government. Most emerging industries begin with few firms as they search for markets and strategies. The number of firms increases as the opportunities in an industry become more apparent and firms compete to establish an industry standard. Once an industry standard has been established and investments and markets are won by a select group of firms, industries begin to consolidate around the strongest and often most efficient firms. Figure 9.1 shows the aggregate growth rate of the number of firms using several illustrative emerging industries, including virtual reality, regenerative medicine, biofuels, and additive manufacturing.[11]

The aggregate annual growth of the number of firms is about 90 percent from roughly the beginning of the emergence of the industries to year five (strategy seekers). The annual growth rate of the number of firms grows to approximately 115 percent over the next five years (standard setters). This annual growth plateaus through year 15 (shakeout leaders). This analysis shows that the growth of the number of firms initially accelerates as industries emerge. However, the growth rate for

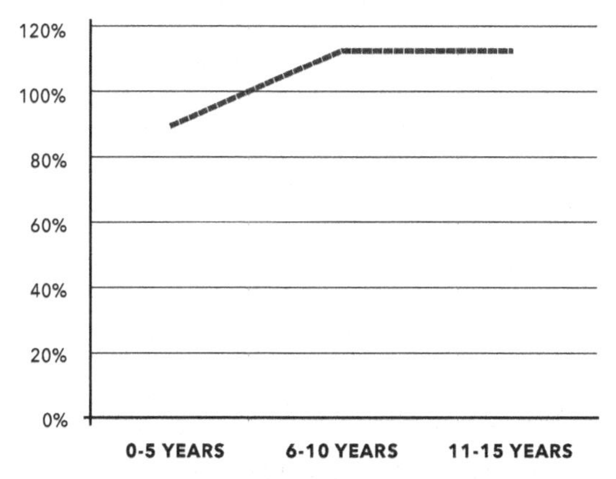

Figure 9.1 Number of Firms (Percentage Growth).

the number of firms levels off over time due to the shakeout of firms that do not possess the industry standard technology, do not win funding from investors, and do not gain market share. In addition, as an industry emerges and there are leading firms, there are likely to be fewer new entrants as entrepreneurs and outside firms perceive entry to be too risky.

9.2.2 Technology

Many emerging industries begin with a discovery or invention, which is often protected by a patent or patents. A patent is the legal right to control the use of a technology that is granted by a government and enforceable within that government's jurisdiction. When technology is further developed, it reaches viability, which is when there is evidence that it accomplishes an intended feat and when a process occurs or a device or substance is produced with the reasonable chance of replication. From viability, a technology may be further developed into a technical solution, which is the deployment or use of a technology to solve a problem.[12] To estimate the development of technology over time in emerging industries, we use the annual growth of granted patents. Figure 9.2 shows the aggregate annual growth of granted US patents for several illustrative emerging industries, including wearable healthcare devices, UAVs, nanotechnology, biofuels, additive manufacturing, virtual reality, and electric vehicles.[13]

The aggregate annual growth rate of granted US patents for several illustrative emerging industries is approximately 150 percent from roughly the beginning of the industries (approximately when an industry enters the Concept state of emergence: technology is discovered,

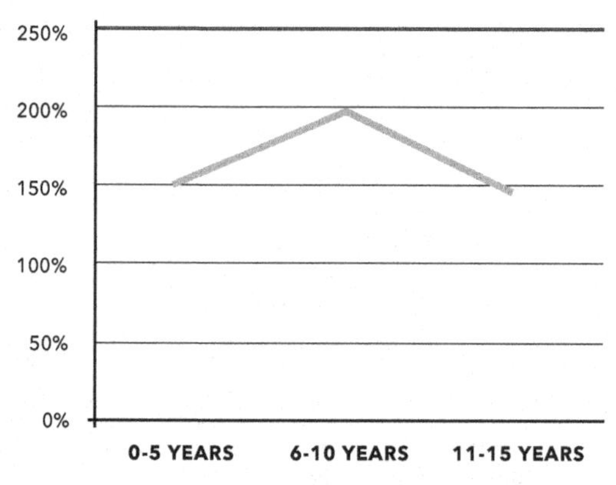

Figure 9.2 Technology – Patenting (Percentage Growth).

visionaries lead markets, firms are bootstrapped and seeking strategies, etc.) to year five, during research and discovery. The annual growth rate of patenting continues to rise to approximately 200 percent over the next five years during the process of becoming a viable technology. However, this annual growth rate falls to about 150 percent from year 11 to year 15 as the technology becomes deployable. This analysis shows that patenting peaks and then declines as industries emerge. This is likely to occur because the limits to discovery of new ideas and inventions are approached and with the emergence of an industry, a technology standard is reached.

9.2.3 *Investment*

Investors are central to emerging industries because they provide money for research, innovation, development of value chains, and commercialization of products and services.[14] As relevant technologies develop and the industry begins to emerge, investment in the form of personal funding, often referred to as bootstrapping, is prominent. Seed funding such as government grants and money from angels and venture capitalists comes next, followed by growth funding such as IPOs and revenue. As an industry emerges, risks decrease or at least are better understood, so more investment is likely to be attracted. Firms seek financing from a number of sources during this period, but due to the necessity of high-risk financing, venture capital firms may be particularly important for the emergence of new industries.[15] The annual aggregate rate of investment is shown in Figure 9.3 using annual investment growth of several illustrative emerging industries, including regenerative medicine, biotechnology, and online video games.[16] The aggregate annual investment

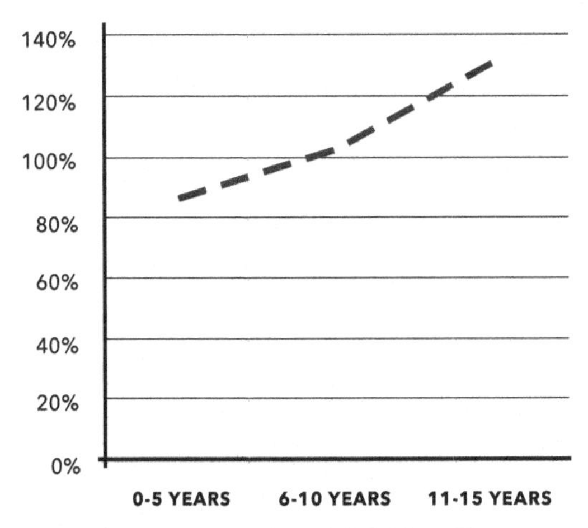

Figure 9.3 Investment (Percentage Growth).

growth is about 90 percent from roughly the beginning of the industries to year five (bootstrapping). The annual investment growth rate grows to approximately 100 percent over the next five years (seed funding). This annual growth continues to grow to about 130 percent from year 11 to year 15 (growth funding). This analysis shows that investment growth accelerates as industries emerge. The differences in aggregate annual growth rates are likely to occur because the early years are dominated by bootstrapping, which is followed by more substantial sources of seed funding from angel and venture capital investors. Even larger growth funding comes next from IPOs and revenue. Investment growth appears to first drive technology growth and then market growth.

9.2.4 Supply Networks

The size of supply networks changes as an industry emerges. Early supply networks are usually sparse because the technology and processes are new and few, if any, suppliers can offer needed technologies, components, and processes. Supply networks grow to meet demand as an industry continues to emerge and are the largest as firms form dense subsectors of expertise to meet the various needs of the new and growing industry. With further emergence, supply networks are likely to concentrate as more successful firms acquire smaller, less healthy firms. This consolidation helps enable the emerging industry to grow, because this leads to process innovation, better consistency of technologies, components, or processes, and lower prices. The price trends for key related technology from illustrative industries is estimated in Figure 9.4. This

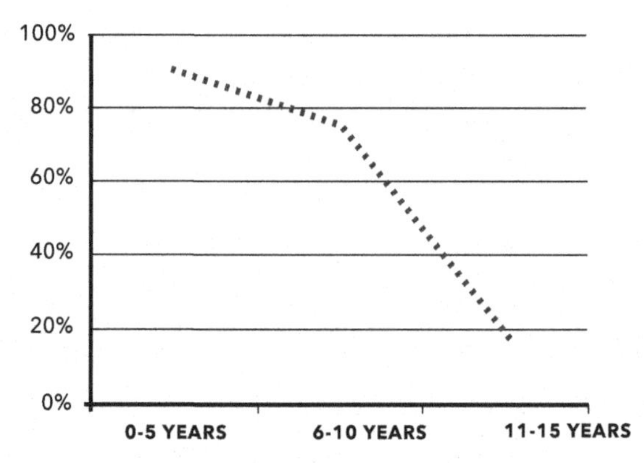

Figure 9.4 Supply Networks (Price Change of Inputs).

average trend line is based on the price of battery technology for electric vehicles and polysilicon for solar photovoltaic panels. The price for electric vehicle batteries and polysilicon falls over 90 percent from roughly the beginning of the industries to year five, approximately 75 percent over the next five years (years 6–10), and approximately 20 percent from year 11 to year 15.[17]

9.2.5 *Production*

The nature of production changes as an industry emerges. Prototype production is often an essential early element of industry emergence, because being able to make something is an important step toward realization of a technology or idea. Pilot production for testing a product is also often essential for proving a concept can be successfully replicated. Scaling production is critical for diffusion, as achieving economies of scale are likely needed to challenge incumbent technologies and industries. Successful production can entice investment, attract market demand, lure new firms to the industry, and motivate advances by supply networks. The scale of production is estimated in Figure 9.5 based on the growth of production in the biofuels and wind power industries as they emerged. Both industries began with prototyping being low scale, pilot production being a significant increase, and scaled production, either batch or continuous, being a much larger scale than prototyping or pilot production.[18]

9.2.6 *Markets*

The development of new markets for technology-based products follows a pattern of visionaries, early adopters, and then followers.[19] The rate of adoption of a technology depends on features of both the technology and

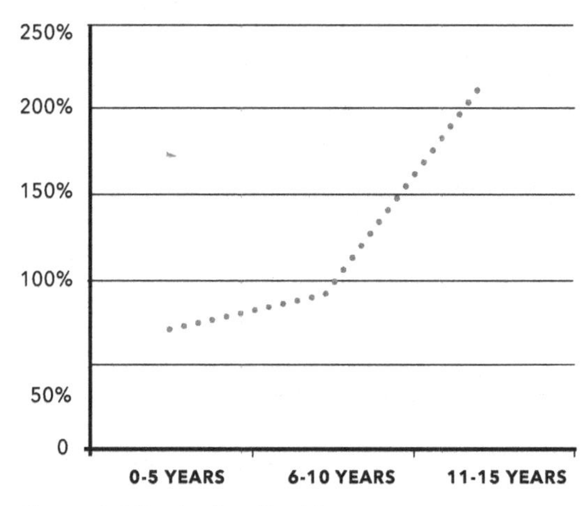

Figure 9.5 Production (Scale).

the potential market. The first customers are visionaries who see and shape the future for the technology by communicating the need and demand for a new product or service solution. Early adopters are next and are keen to adopt new products or services as soon as possible. With further market development, a product or service can diffuse to a much wider audience of followers, as it becomes the standard solution. Figure 9.6 shows market development using aggregate annual sales growth for illustrative emerging industries, including online games, additive manufacturing, and biotechnology.[20]

The aggregate annual sales growth for these illustrative emerging industries is about 120 percent from roughly the beginning of the industries to year five (visionaries). The annual market growth rate grows to

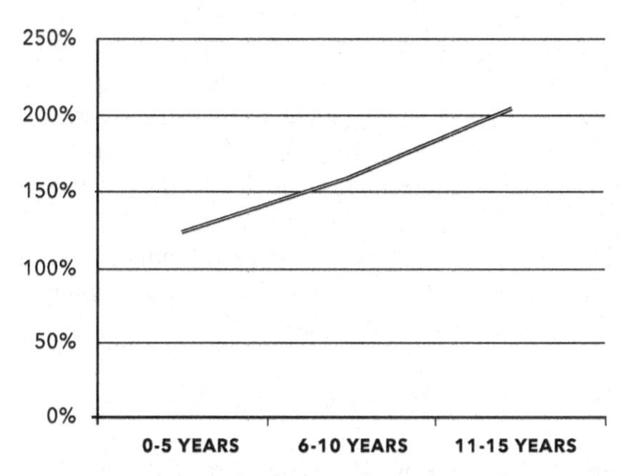

Figure 9.6 Markets (Percentage Growth).

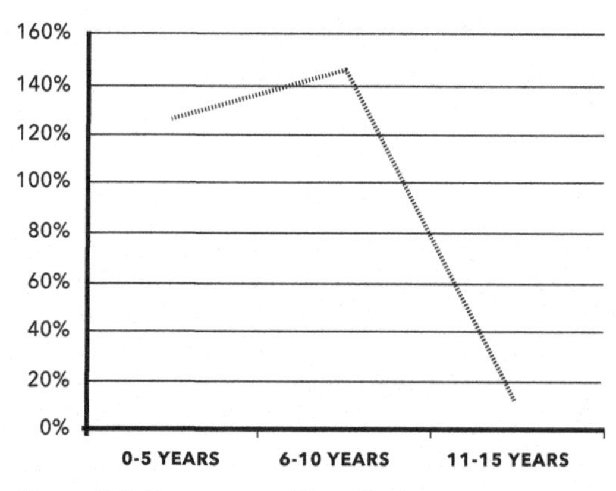

Figure 9.7 Government (Spending).

approximately 150 percent over the next five years (early adopters). This annual growth continues to grow to over 200 percent from year 11 to year 15 (followers). This analysis shows that market growth accelerates as industries emerge. The acceleration of market growth indicates the potential rewards to the firms that develop and protect the leading technologies and market positions.

9.2.7 Government

Government is involved in the birth and growth of emerging industries, providing foundational infrastructure that supports early research, innovation, and technology development.[21] As technologies develop and industries emerge, government grants and procurement become prominent. Governments often invest in early-stage firms, enabling the continued growth of emerging industries. Government involvement in industry emergence continues through the establishment and enforcement of regulation (competition rules, food and drug safety approval, etc.). The amount of government involvement in industry emergence is estimated in Figure 9.7 based on spending for illustrative emerging industries including UAVs and nanotechnology with government-funded infrastructure being the early foundation, followed by increased government involvement in the form of early-stage grant seed funding and procurement, and then declining involvement with the focus on regulation as emerging industries gain momentum.[22]

9.3 Element Synchronization

Synchronization is the coordination of the seven interrelated elements of our industry emergence framework, which we elaborate on throughout the book. Both Figure 1.1 and Figure 9.8 horizontally show the changes for each element that generally occur as an industry emerges. Vertically, we show three states of emergence—Concept, Validation, and Diffusion—which capture the synchronization of the seven elements.

By bringing all of these elements together, Figure 9.8 reveals insights about the nature of interactions occurring between the number of firms, technology, investment, supply networks, production, markets, and government as industries emerge. The growth of the number of firms initially accelerates as industries emerge, but levels off over time due to the shakeout of firms that do not possess the industry standard technology, do not win funding from investors, and/or do not gain market share. A technology standard is established as the market develops, and the strongest firms survive. The industry consolidates, and investment

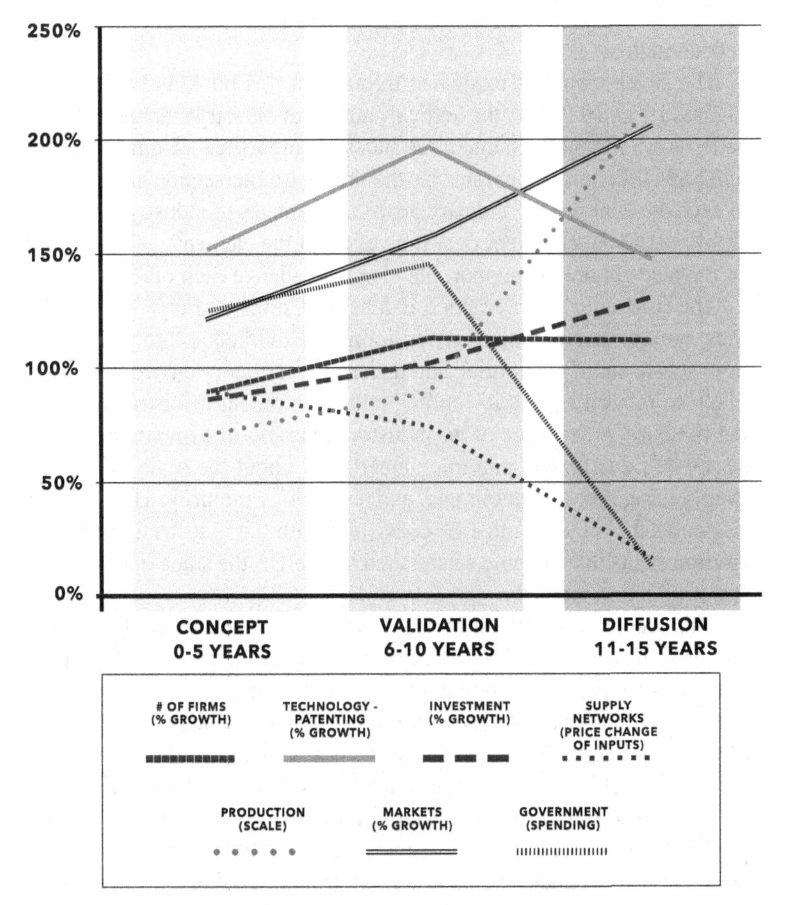

Figure 9.8 Interaction of Elements for Synchronization.

concentrates with the leading firms. The rate of technology development (measured using patent growth rate) peaks and then declines as industries emerge. This is likely to occur because a technology standard is approached and the opportunities for new ideas begin to decline. Investment growth accelerates as industries emerge, driving technology and market growth. Supply networks develop over time and leading suppliers force weaker companies out of business or acquire them by improving their efficiency and lowering input costs. Production builds in volume as industries emerge from prototypes, to pilot testing, to scaled production. Market growth (sales) accelerates as industries emerge and indicates the potential rewards to the firms that develop and protect the leading technologies. Government involvement in industry emergence initially grows through infrastructure provision, seed funding, and procurement, only to taper off and decline as private investment and firm strategy gain prominence and as government shifts its focus to regulation and standards.

The seven elements explained throughout this book and exhibited in Figures 1.1 and 9.8 show the vertical patterns of industry emergence. These patterns are summarized into three states of emergence—Concept, Validation and Diffusion—to emphasize the timing of intersection and deviation of each element driving industry growth. The states of industry emergence highlight the development over time of all of the elements, which have to be in place not only to support but also to challenge each other.

The first state of emergence is labeled **Concept.** Early stage industries, with technology discoveries, immature markets, prototype production, sparse supply networks, and uncertain regulatory environments, are likely to rely on bootstrap funding and government infrastructure. During the early emergence of an industry, a variety of firm strategies may be applied due to, for example, uncertainty about the scale of demand, supply base, investment climate, and technology maturity. The dominant strategies that emerge must be consistent with the industrial and market environments that become established, as well as the stage of technology and production maturity.

The second state of emergence is labeled **Validation**. Technology has proven to be a viable solution to a market need of early adopters. Advances in technology and markets attract seed funding, and standard-setting firms begin to take over industry leadership. The lack of a network of dedicated suppliers as an industry first emerges leads to a growing group of collaborating new firms engaging in a wide range of value chain activities resulting in declining input costs. As demand for a successful product increases, the supply network and production must develop if large volume production is to be achieved and sustained.

Diffusion is the third state of emergence. Industries with proven technology, growing demand from market followers, reduced risk, and standard-setting are likely to rely on the presence of larger-scale investors, IPOs, or self-generated revenue to enable the establishment of growing production capabilities, the refinement of products and technologies, and the development of supply networks capable of delivering the required process improvements.

9.4 Conclusion

A central tenet of our book is that the elements we highlight are influenced by each other in many ways, and they co-develop and need to synchronize in order for an industry to emerge and grow. The development and change of the elements affecting an emerging industry take place in parallel. The elements affect each other as they interact and develop, in some cases driving each other forward and in other cases restricting each other.

This chapter and especially Figure 9.8 illustrate the opportunities and the necessity for synchronization of the elements for an industry to emerge. While it is valuable to understand the elements influencing emerging industries and their interactions, it is essential to understand when and how synchronization of the elements can be achieved, as this enables planning and speculation for inventors, firms, investors, and policymakers.

Exercises

- Choose an emerging industry and assess the interaction between the elements affecting the industry.
- Choose an emerging industry and assess the degree of synchronization of the elements of the industry.
- Identify policies or approaches for aligning elements that may be misaligned in order for an industry to emerge.

Notes

1 Teece, D. J. (1986). Profiting from technological innovation: Implications for integration, collaboration, licensing and public policy. *Research Policy*, *15*(6), 285–305.
2 Rothaermel, F., & Hill, C. (2005). Technological discontinuities and complementary assets: A longitudinal study of industry and firm performance. *Organization Science*, *16*(1), 52–70.

3 Klepper, S., & Graddy, E. (1990). The evolution of new industries and the determinants of market structure. *The RAND Journal of Economics, 21*(1), 27–44.
4 Rogers, E. (2003). *Diffusion of innovation.* New York, NY: Simon & Schuster International.
5 Reed, F., & Walsh, K. (2002). Enhancing technological capability through supplier development: A study of the UK aerospace industry. *IEEE Transactions on Engineering Management, 49*(3), 231–242.
6 Jolly, V. (1997). *Commercializing new Technologies: Getting from Mind to Market.* Boston, MA: Harvard Business School Press.
7 Dodgson, M. (2000). *The Management of Technological Innovation: An International and Strategic Approach.* Oxford, UK: Oxford University Press.
8 Adner, R., & Levinthal, D. (2002). The emergence of emerging technologies. *California Management Review, 45*(1), 50–66.
9 Bradford, T. (2007). *The Solar Revolution: The Economic Transformation of the Global Energy Industry.* Cambridge, MA: MIT Press.
10 The data analysis in the firm strategy, technology, investment, supply networks, production, market, and government sections is based on data from approximately the first, fifth, tenth, and fifteenth years of emergence of several illustrative industries identified in each section. The curves are based on the average change between these data points.
11 Available online at http://frenchtechhub.com/blog/2015/09/whats-actually-happening-in-the-virtual-reality-vr-industry/ (accessed February 2, 2016); Greenlight, Virtual Reality Estimates. Available online at http://venturebeat.com/2015/10/12/the-landscape-of-vr-is-complicated-with-234-companies-valued-at-13b/ (accessed February 2, 2016); Wohlers Reports 2013, 2014, and 2015, Additive Manufacturing; Renewable Fuels Association (2015). Available online at www.ethanolrfa.org/resources/industry/statistics/#EIO (accessed May 7, 2015); Alliance for Regenerative Medicine (2016), State of the Industry Briefing. Available online at http://alliancerm.org/event/regenerative-medicine-state-industry-briefing (accessed January 11, 2016); Nerem, R. (2010). Regenerative medicine: The emergence of an industry. *Interface*, 7, 771–775.
12 Dodgson, M. (2000). *The Management of Technological Innovation: An International and Strategic Approach.* Oxford, UK: Oxford University Press.
13 Available online at https://sites.google.com/site/analyzingpatenttrends/case-study-hybrid-electric-vehicle-technology (accessed February 2, 2016); Greenbaum, E. (2014). A virtual reality patent landscape analysis. Available online at https://greenbaumpatent.wordpress.com/2014/01/23/a-virtual-reality-patent-landscape-analysis/ (accessed February 2, 2016); D'Aveni, R. (2015). The 3D printing revolution. *Harvard Business Review*; Yan Dang, Yulei Zhang, Li Fan, Hsinchun C., & Roco, M. (2010) Trends in worldwide nanotechnology patent applications: 1991 to 2008. Journal Nanoparticle Research *12* (3), 687–706; Available online at www.researchgate.net/figure/277846293_fig2_Figure-4-Biofuel-patent-families-transnational-patents-in-the-sample-countries-Source (accessed February 2, 2016); Carrascosa, S., & Escorsa, E. (2014). Hovering over the drone patent landscape, *IALE Tecnologia*; Available online at www.advocaterahuldev.com/health-fitness-tracking-patents-devices/ (accessed February 2, 2016).
14 Lamoreaux, N. R., & Sokoloff, K. L. (Eds.) (2007). *Financing Innovation in the United States: 1870 to the Present.* Cambridge, MA: MIT Press.

15 Von Burg, U., & Kenney, M. (2000). Venture capital and the birth of the local area networking industry. *Research Policy*, *29*, 1135–1155.
16 Global Video Games Investment: China, Online, Mobile Ascendent, by Tim Merel. Available online at www.gamasutra.com/view/feature/6294/global_video_games_investment_.php?print=1 (accessed February 2, 2016); Money Tree TM (2014). *Biotechnology investments*. Available online at www.pwcmoneytree.com (accessed February 2, 2016); Available online at www.gamasutra.com/view/news/264873/Total_value_of_game_industry (accessed January 11, 2016); Alliance for Regenerative Medicine (2016) State of the Industry Briefing; Nerem, R. (2010). Regenerative medicine: The emergence of an industry. *Interface*, *7*, 771–775.
17 Farrell, J. (2012). Are the batteries ready? 100% clean energy requires progress on storage. Available online at http://grist.org/article/are-the-batteries-ready-100-clean-energy-requires-progress-on-storage/ (accessed February 2, 2016); Available online at www.thequartzcorp.com/en/blog/2014/11/03/polysilicon-demand-and-the-solar-industry/100 (accessed February 2, 2016).
18 U.S. Energy Information Administration (EIA) International Energy Statistics; OECD-FAO Agricultural Outlook 2014–2023. OECD. Available online at http://stats.oecd.org/view (accessed May 7, 2015).
19 Rogers, E. (2003). *Diffusion of Innovation*. New York, NY: Simon & Schuster International.
20 Wohlers Reports 2013, 2014, and 2015, Additive Manufacturing; Available online at www.statista.com/statistics/270728/market-volume-of-online-gaming-worldwide (accessed February 2, 2016); www.bbbiotech.ch/en/bb-biotech/; Available online at www.ibisworld.com/industry/global/global-biotechnology.html (accessed February 2, 2016).
21 Block, F., & Keller, M. (Eds.) (2010). *State of Innovation: The U.S. Government's Role in Technology Development*. Boulder, CO: Paradigm Publishers.
22 Tesla, C. (2014). Shaping the future atom by atom – An introduction into the nanotechnology industry and its future Available online at www.tumotech.com/2014/06/09/shaping-the-future-atom-by-atom-a-brief-introduction-into-nanotechnology/ (accessed February 2, 2016); Waldman, P. (2013). Games of drones. *The American Prospect*.

Further Reading

Malerba, F. (2006). Innovation and the evolution of industries. *Journal of Evolutionary Economics*, *16*, 3–23.

Moore, G. (2002). *Crossing the Chasm: Marketing and Selling High-tech Products to Mainstream Customers*. New York, NY: Harper Publishing.

Nelson, R. (1994). Co-evolution of technology, industrial structure and supporting institutions. *Industrial and Corporate Change*, *3*(1), 47–63.

10 Strategic Management for Industry Emergence

Learning Objectives:

- Review the key elements of the emerging industries framework
- Assess the implications of understanding industry emergence

Key Concepts:

- Drivers of industry emergence
- Implications of industry emergence

10.1 Introduction

The study of industry emergence is challenging but important because of the potential for technological advancement, improvements to society, firm and job growth, and economic development. However, understanding emerging industries has been limited in large part due to focus on single industries, using a single disciplinary approach, and/or considering a single perspective. Little has been done to understand the systemic nature of emerging industries. This book presents an interdisciplinary framework of elements that exhibits the interconnection, coevolution, and synchronization of the elements that underlie industry emergence. The seven elements we focus on are unlikely to be the only ones that influence industry emergence. Other elements may be critical for a particular industry, and the significance of some, government, for example, will vary significantly from one industry to another. More generally, the art of the analysis lies in understanding which elements are most relevant for a given industry at a particular point in time, and how these elements are linked, change over time, and need to be synchronized in order for an industry to emerge.

10.2 Implications of Understanding Industry Emergence

Understanding how industries emerge is informative for inventors, firms, investors, scholars, and policymakers. Inventors and firms can benefit from a better understanding of the multiple elements influencing the opportunities for, and progress of, their technology. Investors can make better-informed decisions about their funding of companies, scholars can present their deep but narrow findings in a more systemic context, and policymakers can coordinate their efforts in concert with firm strategy, technology, investment, supply networks, production, and markets and seek to encourage synchronization between the elements.

10.2.1 Inventors and Firms

For inventors and firms, this book draws attention to the need to consider and attempt to influence the synchronization of a range of elements as they seek to commercialize a technology or business idea. Inventors and managers of firms can fall prey to focusing on too few elements when projecting the evolution of their technology and business. They may envision success because they have a leading technology, only to be frustrated by delays and failure due to unforeseen complications with other elements. This book stresses the need for a broader and deeper understanding: one that includes the elements presented in our framework, but also one that seeks to anticipate the connections between the elements and takes a proactive approach to influence these aspects by educating investors and policymakers, considering new business models, and building supply networks and production capabilities.

In addition, our analysis of numerous emerging industries led us to the conclusion that it is more challenging to start a company based on a new technology after a technology standard is set. Once the patenting rate peaked and started to fall, the rate of firm entry plateaued and even declined in some industries. Therefore, inventors and firms need to advance their technology or business idea quickly, gain intellectual property protection, and build market support to become the industry standard. Industry entry and gaining market share are more difficult once industry consolidation around a technology standard begins. If technology and market leadership are not possible, inventors and firms can look for alternative applications in different emerging industries or possibly build a supplier role providing components or services to the industry leaders.

10.2.2 Investors

Investors are likely to gain valuable insights from this book, and they may choose new investment strategies based on a more complete picture of industry emergence. They will need to weigh the risks of many aspects of an industry and its context, which will offer them a more realistic picture of the situation they face. Investors can use our framework of elements affecting emerging industries and our insight on element synchronization to rate the state of emergence of an industry and analyze the likely hurdles for company and industry growth.

Technology due diligence entails an assessment of a company's and its peers' development of the technology necessary for the emerging industry to move forward. This process involves a review of the technology and the companies' ability to develop it into a feasible product. With the help of industry experts, investors can assess the development of the technology of its targeted company and other companies in the industry. This can help provide insight on the likely time needed to have a working product and the funding needed for the technology to mature to the level of operability. Both the time and funding elements are necessary to make judgments about the probability of technology advancement, which the success of the new venture often relies upon.

Closely related to technical feasibility is intellectual property protection related to the technology, products, or processes. An investor's assessment usually needs to include a review of the status of patents or other protection of intellectual property (issued and pending patents), a review of competing technologies, and a risk assessment of the potential for there to be challenges to the protection of intellectual property. Investors are likely to face lower risk after a technology standard is set and patenting declines, but there may also be fewer rewards.

If the emerging industry involves production, a review of the state of production is essential. This would entail an assessment of the production processes for the industry's products, an understanding of how successful companies have been with prototype development and how many and what type of production facilities (demonstration and commercial facilities) are currently in use. It is also useful to understand the state of the emerging industry's supply networks. It is often important to know the development of the suppliers of key inputs to an emerging industry and whether supply networks are likely to be a bottleneck to the emergence of the industry.

Investors can also benefit from this book by using the insight to assess the market potential of companies and their emerging industries. An emerging industry needs to address a challenge or problem that potential customers face. Being able to address this challenge or problem can represent a market opportunity, and part of the review process is to evaluate whether the industry's technology and product could solve an important challenge or problem for visionaries, early adopters, and followers.

Investors also need to size the market and its geographic reach. These aspects matter for assessing the projected revenue (size of the market) and the likely marketing and distribution costs due to the vastness of the estimated market. This process involves estimating the size and location of the potential customer base and the percentage (or market share) of this market the product will be able to attract. This review involves an assessment of competing technologies, the existing market, the potential for market growth, and the likely competitive advantage the new industry's product may hold.

Investors should also identify the events and factors upon which the success of the industry's companies depends. This is often referred to as risk identification because it involves identifying what and who could stand in the way of the success of the companies. This involves identifying the events that need to occur for the industry's emergence and estimating the consequences of each event failing to occur in support of the industry.

10.2.3 Scholars

This book presents a broad and systemic view of industry emergence. This presentation offers scholars a foundation to build upon and test. Future research needs to delve more deeply into each of the elements we present, the patterns of relationships between the elements, and the behaviors of entities observed in the emergence of new industries. These elements and patterns appear to be far from random, but they are also not fixed or predetermined. The ways in which the elements synchronize and the way entities mobilize to bridge the gaps in the emergence of an industry provide scope for future research that would advance our ability to understand and nurture the development of new industries.

10.2.4 Policymakers

Policymakers can benefit from the insight in this book by focusing their attention on emerging industry elements and their interaction for synchronization. They can begin with an identification of the critical economic conditions, regulations, subsidies, and tax benefits upon which the emergence of a new industry is likely to be based. Policymakers should then assess the range of applications of each of these for the industry to emerge and estimate the effects on the industry's companies' projected performance. The time frame for variation of these circumstances should be specific to the business ventures and technology, and the variations need to be included in the financial analysis as well. For example, it may be optimal for government to reduce its investment in an emerging industry once private investment gains momentum, thus enabling investment in other earlier-stage emerging industries.

Policymakers can encourage the emergence of particular industries by coordinating the multiple parties representing the elements most relevant to their targeted industries and by encouraging appropriate interaction and synchronization of key elements in order to help an industry emerge and continue its growth. Some elements may lag behind others, causing an industry to grow slowly or even contract. The approach covered in this book can help policymakers identify actions to better align the development of the elements affecting the emergence of industries.

10.3 Conclusion

This book contributes to our understanding of how industries emerge as they undergo three states of emergence based on seven interconnected elements. Furthermore, it explains the importance of element synchronization within each state of emergence to enable progress in the emergence of industries. The framework presented in the book is based on the evidence from case studies and quantitative data, which can serve as a foundation for additional testing of the framework and the elements on different types of industries, industries in different contexts, and industries at different states of emergence. We envision this book helping multiple parties as they play their roles in the emergence of an industry, but we also seek to continue the conversation about how to better understand and encourage the emergence of new industries.

Exercises

- Take the perspective of an inventor in an emerging industry and assess how the seven elements are likely to affect this inventor's success.
- Describe how policymakers may be able to influence the emergence of an industry.
- Identify additional elements other than the seven in this book that may affect industry emergence.

Further Reading

Block, F., & Keller, M. (Eds.) (2010). *State of Innovation: The U.S. Government's Role in Technology Development*. Boulder, CO: Paradigm Publishers.

Funk, J. (2012). *Technology Change and the Rise of New Industries*. Palo Alto, CA: Stanford University Press.

Lamoreaux, N. R., & Sokoloff, K. L. (Eds.) (2007). *Financing Innovation in the United States: 1870 to the Present*. Cambridge, MA: MIT Press.

Index

additive manufacturing 9, 55, 57–58, 86–87, 91
alliance gap 46
angel investors 7, 33–34, 36–38, 64, 88–89
application gap 24, 29

biofuels 39–40, 67, 80, 86–87, 90
biotechnology industry 39, 43, 59, 77, 80–82, 91
bootstrap funding 6, 19, 32–33, 35–38, 40, 88–89, 94

collaborating suppliers 44–46
concept state of emergence 4, 9, 19, 26, 30, 35, 58–59, 84, 87, 93–94
corporate investment 29, 33, 35

deployment 6, 23–24, 28, 30, 87
diffusion state of emergence 9, 19, 26, 39, 58–59, 76, 84, 93–95

early adopters 8, 16, 29, 47, 59, 62–67, 70–71, 77, 79, 85, 90–92, 94
electric vehicles 39, 47–48, 63–65, 80, 90
electronic commerce industry 66–67
emerging industries 1
expertise gap 45, 47

firm 11
firm strategy 6, 11–12, 86–87
followers 8, 16, 62–63, 66–67, 70–71, 85, 90–92, 95, 100

government 8, 74–75, 92
growth funding 32, 33, 35, 38–40, 88–89

industry 2
industry emergence 4
industry emergence framework 5
industry shakeout 3, 6, 11–12, 16–19
infrastructure 74–78, 94
intellectual property 8, 12, 55, 59, 75–76, 78, 81, 100
inventors 99
investment 6, 32–33, 88–89
investors 99–101

knowledge gap 23, 29, 85

lead suppliers 45–46, 49–50
legacy suppliers 43–49

market expansion gap 67–70
market interest gap 63–65, 67–69
markets 8, 62–63, 90–92
milestone gap 35

nanotechnology 55, 59, 87, 92

online video game industry 17–19, 88
operability gap 55
operational effectiveness gap 16, 19

piloting 54–56
policy gap 76
policymakers 101–102
positioning gap 13, 15, 16, 19
process gap 77
procurement 8, 26, 67, 70, 74, 76–77, 79–80, 82, 85, 92
production 7, 53–54, 90
prototyping 7, 53–55, 57–58, 90

regenerative medicine industry 24, 28–30, 57, 88
regulation 8, 74, 77, 79–82, 85, 92, 94, 101
research and discovery 23, 28–30, 88
risk–reward 33–34, 38–40
robotics 58

scaling 3, 7, 16, 40, 53–54, 56–57, 60, 67, 90
seed funding 19, 32–36, 38–40, 59, 76, 88–89, 92, 94
shakeout leaders 11–13, 16, 86
social networking industry 36–40, 77
solar photovoltaic (PV) industry 64, 68–71
standard setters 11–12, 15–16
strategy seekers 12–13, 86
supply networks 7, 43–44, 89–90
synchronization 1, 84, 92–94

technology 6, 22–23, 87–88
technology contingency 24, 29–30
technology readiness levels 27–29

unmanned aerial vehicles (UAV) 13–15, 65, 77, 87, 92

validation state of emergence 4, 9, 19, 26, 30, 39, 58–59, 84, 93–94
venture capital 7, 26, 29, 33–35, 38, 47–48, 88–89
viability 6, 23–24, 28–30, 34, 59, 69, 77, 87
virtual reality industry 78–79, 86
visionaries 8, 62–64, 66, 68, 71, 85, 88, 90–91, 100
volume gap 56, 58

wearable healthcare devices 25–27, 87
wind turbine industry 49–50, 90